# INSIDE THE MIND OF AI

KIM HAZEL

WILLIAM HASELTINE

ACCESS
HEALTH INTERNATIONAL

www.williamhaseltine.com



Cover art by Kim Hazel, PhD



All author proceeds from the sale of this book will be donated to the nonprofit global think tank ACCESS Health International.

*Some narratives have been shortened for clarity. Please note that names and identifying details may have been altered to protect privacy, and certain events may have been condensed.*

# ACKNOWLEDGMENTS

This work is supported by ACCESS Health International (www.accessh.org).

---

**To Dr. William Haseltine**

There is no version of this book that exists without you. The clarity of its science, the honesty of its questions, and any wisdom the reader finds in these pages belong, in some measure, to the years you have spent teaching me how to think about biology, medicine, and the responsibilities that come with writing about either one.

Working with you on this book, as on the projects that came before it, has been one of the highlights of my professional life. Thank you for trusting me with your ideas, for lending this project the weight of your name, and for the mentorship you've given me.

✧✧✧

*Kim Hazel, Ph.D.*

*To my loving husband, Colin Purcell.*
*To my tech-loving friends: Emily Trout, John Andrick, Wes & Audrey Coolbaugh.*

✧✧✧

# BOOKS BY WILLIAM A. HASELTINE

***Affordable Excellence:*** *The Singapore Healthcare Story* (2013)

***Improving the Health of Mother and Child***: *Solutions from India* (2014)

***Modern Aging***: *A Practical Guide for Developers, Entrepreneurs, and Startups in the Silver Market* (2015)

***Aging with Dignity***: *Innovation and Challenge is Sweden-The Voice of Care Professionals* (2017)

***Every Second Counts***: *Saving Two Million Lives. India's Emergency Response System.The EMRI Story*(2017)

***Voices in Dementia Care*** (2018)

***Aging Well*** (2019)

***World Class***. *Adversity, Transformation, and Success, and NYU Langone Health* (2019)

***A Family Guide to Covid***: *Questions and Answers for Parents, Grandparents, and Children* (2020)

***A Covid Back To School Guide***: *Questions and Answers for Parents and Students* (2020)

***Covid Commentaries***: *A Chronicle of a Plague, Volumes I, II, III, IV, V, and VI* (2020)

***My Lifelong Fight Against Disease***: *From Polio and AIDS to Covid-19* (2020)

***Science as a Superpower***: *My Lifelong Fight Against Disease And The Heroes Who Made It Possible* (2021)

***Variants!***: *The Shape-Shifting Challenge of Covid-19 Vaccine Evasion & Reinfection* (2021)

***COVID-Related Post-traumatic Stress Disorder (CV-PTSD)*** (2021)

***Natural Immunity And Covid-19***: *What It Is And How It Can Save Your Life* (2022)

***Omicron***: *From Pandemic to Endemic* (2022)

***Monoclonal Antibodies:*** *The Once and Future Cure for Covid-19* (2023)

***The Future of Medicine***: *Healing Yourself: Regenerative Medicine Part One* (2023)

***Viroids and Virusoids***: *Nature's Own mRNAs* (2023)

***CAR T***: *A New Cure for Cancer, Autoimmune and Inherited Disease* (2023)

***Ending Hepatitis C***: *A Seven-step Plan for a Successful Eradication Program* (2023)

***Better Eyesight***: *What You and Modern Medicine Can Do to Improve Your Vision* (2024)

***Fusion!*** *The Melding of Human and Machine Intelligence* (2024)

***Checkpoint Inhibitors***: *A Key to Curing Cancer* (2025)

***Live Longer***: *What You Can Do, What Medicine Can Do* (2025)

***Destiny's Child No Longer:*** *Rewriting Genetic Fate* (2025)

***Biology of Desire:*** *Your Brain & Behavior* (2026)

✧✧✧

***Recent Textbooks***

***The COVID-19 Textbook***: *Science, Medicine and Public Health* (2023)

***Molecular Biology of SARS-CoV-2***: *Opportunities for Antivirus Drug Development* (2024)

***The Biology of RNA*** *Volume 1* (2026)

***The Biology of RNA*** *Volume 2* (2026)

✧✧✧

CHAPTER 1

# HOW MACHINE INTELLIGENCE WORKS & HOW IT FAILS

YOU HOLD a plate with a ripe tomato, a crisp carrot, and a lumpy potato. With a quick flick of your wrist, you turn the plate upside down.

You don't even need to see what happens next. The vegetables tumble through the air and smack onto the floor, just as you pictured. You didn't reach for equations or crunch probabilities. Instantly, your mind played out a vivid prediction of what came next.

Ask one of the systems behind today's machine intelligence the same question, and a strange thing happens. It often gets the answer wrong. It picks a response that "sounds" right in a sentence but cannot be true in a kitchen. The model is not glitching. It is doing exactly what it was built to do. It models language. It does not model the world. That single mismatch sits at the center of everything that follows.

## BEFORE THE MACHINES

The story of thinking machines starts long before the first wire was soldered. In the 17th century, two of Europe's restless minds reached for arithmetic with brass and wood. Blaise Pascal, a French prodigy who built one of the earliest mechanical calculators to ease his father's tax work. Gottfried Wilhelm Leibniz, a German philosopher who designed a machine that could multiply by repeated addition. Their contraptions were beautiful, expensive, and fussy. They did sums. Little more.

Two centuries later, an English mathematician named Charles Babbage drew up something much grander. His Difference Engine was supposed to print perfect mathematical tables, the kind used by navigators and engineers. The full design called for 25,000 brass parts and four tons of metal. He built a working slice of about 2,000 parts. The British government, after years of mounting bills, walked away.

Babbage kept drawing. His next design, the Analytical Engine, was a machine with two halves: a "store" that held numbers, and a "mill" that operated on them. Memory and processor, in Victorian dress. Instructions and data would arrive on punched cards borrowed from the Jacquard loom, the weaving machine that could turn a stack of cards into a tapestry. Change the cards, change the cloth. Change the cards, change the calculation. The Analytical Engine was never built. The idea outlived the iron.

In 1843, an English mathematician named Ada Lovelace translated a paper by the Italian engineer Luigi Menabrea on Babbage's machine and added notes that were nearly three times as long as the original. In a passage labeled Note G, she described how the engine could compute Bernoulli numbers, a famously prickly sequence that turns up in calculus and number theory. Many historians call this the first computer program.

Lovelace saw further than her teacher. The engine, she wrote, was not just an arithmetic machine. Give it the right symbolic language, and it could manipulate anything that could be symbolized. Music, for example. She imagined a machine composing. It would take more than a century for that idea to land.

Late in the 19th century, an American inventor named Herman Hollerith put punched cards to work on a more practical problem:

the 1890 U.S. Census. His tabulating machine read holes in cardboard, counted, and sorted. A task that would have taken years finished in 30 months across 62 million records. The company he founded eventually became IBM.

These machines were marvels of focus. Each one was built for a single job. None could be reprogrammed by changing its mind, only by changing its body. The dream of a single, general-purpose, programmable machine stayed on paper.

## TURING'S UNIVERSAL MACHINE

The breakthrough came from a question. In 1936, a young British mathematician named Alan Turing published a paper with a forbidding title and a simple idea inside. He asked whether there could be a mechanical procedure that decides, for any mathematical statement, whether it is true or false.

To attack the question, he invented an imaginary device. Picture a paper tape stretching to the horizon, divided into squares. A head sits over one square at a time. Rules tell it what to do. If you see this symbol in this state, write that one, move one square, switch states. That is the whole machine.

Then Turing made the move that changed everything. He showed that a single such machine, given the right instructions, could imitate any other. A universal machine. One device, any program.

The paper was abstract and dense. What it gave the world was a clean definition of what it means for a process to be computable and a proof that a single general machine could handle all of them. Babbage's vision, rewritten in the language of logic.

## WAR & THE BIRTH OF ELECTRONIC COMPUTING

Turing's idea might have stayed on the shelf for another generation if the world had not gone to war. The Second World War created a hunger for calculation that no human hand could feed. Codes had to be broken. Shells had to be aimed. Bombs had to be designed. The work was urgent, the money flowed, and the rules about what could be tried loosened.

At a country estate north of London called Bletchley Park, a strange community of mathematicians, chess players, linguists, and engineers set to work on the German cipher machine, Enigma. Enigma looked like a typewriter with rotors that clicked one step every time a key was pressed. Each click reshuffled the wiring inside. Two operators with the same starting setup could type at one machine and read at the other.

The Germans considered Enigma uncrackable. The numbers were on their side. The basic Enigma had five rotors to choose from and a plugboard that swapped pairs of letters. The possibilities multiplied into a staggering 103 sextillion combinations. If you tried a new setting every second, you'd run out of time long before you ran out of options—the sun would burn out, and you'd still be guessing.

Turing and his colleagues found a way in. They leaned on the messiness of real war traffic. Weather reports came in at the same time each morning. Operators reused old keys. Standard greetings and sign-offs repeated. From those cracks, Turing and the engineer Gordon Welchman designed a machine called the Bombe. Built of clattering relays and drums, the Bombe rifled through possible Enigma settings, throwing out the ones that contradicted a guessed phrase. By early 1940, the British were reading German radio. By the end of the war, they were decoding about one in ten Enigma messages, and the intelligence was shaping the war.

The Germans shielded their most sensitive secrets with an even more complex cipher: the Lorenz cipher. Ordinary relays couldn't keep up with its labyrinth of scrambled signals. The British answer was Colossus. It was a machine unlike any before it. Tommy Flowers, a quiet Post Office engineer, marshaled his team to build this beast of wires and glowing vacuum tubes. Colossus didn't just calculate. It processed code at electric speed, outpacing the whir of gears and the patience of men. By war's end, ten of these computers hummed in secret rooms, flickering as they unraveled enemy messages.

Colossus was programmable in the sense that operators could change its function by rewiring plug panels and setting switches. It was designed for one specialized task: codebreaking. It was not a general-purpose computer in the modern sense, and its existence remained secret for decades.

## THE BEAST OF A BRAIN

Across the Atlantic, American engineers were building their own electronic giant. In 1942, the physicist John Mauchly proposed an all-electronic calculator to the U.S. Army. The Army needed firing tables, the dense charts that told gunners how to aim under different wind, range, and temperature. Human computers using mechanical calculators took weeks per table. Battles could not wait weeks. Mauchly partnered with the engineer J. Presper Eckert. Construction began in 1943. In 1946, the result rolled to life: the Electronic Numerical Integrator and Computer. ENIAC.

ENIAC was a beast. 1,800 square feet of floor. 30 tons. 200 kilowatts of power, enough to dim the lights in the surrounding neighborhood when it powered on. It also contained nearly 18,000 vacuum tubes, 1,500 relays, and 70,000 resistors. When a tube burned out, which happened frequently, technicians had to find and replace it before the machine could continue. Despite this, ENIAC was also fast.

It performed 5,000 additions per second, 1,000 times the pace of the machines it replaced. Its first job was a calculation for the hydrogen bomb that would have taken a roomful of human computers a year. ENIAC finished in hours. For ten years it may have run more arithmetic than the entire human species had run in all the centuries before.

ENIAC was also programmable. However, programming meant pulling cables and resetting switches across panels the size of a wall.

Each new problem meant essentially rebuilding the machine. Six women are credited with doing this work. Kay McNulty, Betty Jennings, Betty Snyder, Marlyn Wescoff, Fran Bilas, and Ruth Lichterman. They were handed logical diagrams with no manuals. Yet they learned how the machine worked from the inside out.

The women figured out from circuit traces how the machine actually thought. They invented programming and debugging techniques that form the foundation of computer science. Today, they are celebrated as among the very first computer programmers in history.

## THE STORED-PROGRAM REVOLUTION

The breakthrough that transformed computing from specialized calculating engines into true general-purpose machines came from a deceptively simple idea: store the program in the computer's memory, just like data.

In 1945, the mathematician John von Neumann, by then working with the ENIAC team, wrote a report on the next machine in line. He documented this concept and named it the Electronic Discrete Variable Automatic Computer. He argued that instructions and numbers should be stored together in a single electronic memory. The idea was not entirely his alone. Babbage's analytical engine used cards for both, and Turing's abstract machine treated rules and symbols as a single stream. Von Neumann, however, gave the principle its clean modern form.

The payoff was huge. A stored-program computer did not need to be rewired for each job. Programs could be loaded, edited, even written by other programs. The machine could finally live up to Turing's promise of universality.

The first machine to pull this off was a small experimental device at the University of Manchester nicknamed "the Baby". Built by Frederic Williams, Tom Kilburn, and Geoff Tootill, the Baby ran its first stored program on June 21, 1948. It had just 1,024 bits of memory, barely enough to hold 32 numbers. It was tiny by comparison to the beast that was the ENIAC. Most importantly, it proved the concept worked. Programs and data lived together in electronic memory, retrieved and executed at electronic speed.

By the early 1950s, the age of gears and plugboards was coming to an end. The machines were still huge, still costly, still cranky, and still confined to universities and ministries. The architecture, however, was in place. Within a generation, vacuum tubes would shrink to transistors, transistors to chips, chips to slivers of silicon thinner than a fingernail. The universal machine Turing imagined in 1936 would end up in your pocket.

As the hardware evolved from clattering mechanisms to silent silicon, a new frontier emerged in the logic that guided them. The question shifted: *what kinds of instructions, rules, and patterns could these machines follow? If their bodies were now versatile, could their minds be made flexible too?*

## THE RULE-BASED APPROACH

As computers and concepts of machine intelligence developed together through the 1940s and 1950s, medicine was among the first fields to see their potential. The logic seemed sound. *If diagnosis were a matter of matching symptoms to diseases according to known rules, couldn't a machine do it faster and more reliably than a human?*

In 1959, physician Keeve Brodman and his colleagues at Cornell University argued that "the making of correct diagnostic interpretations of symptoms can be a process in all aspects logical and so completely defined that it can be carried out by a machine." The statement was bold, even provocative, suggesting that one of medicine's most cherished skills—clinical judgment—might be reducible to computation.

Eleven years later, William B. Schwartz wrote in the *New England Journal of Medicine* that computing science would likely augment, and in some cases largely replace, the intellectual functions of physicians. He predicted confidently that by the year 2000, computers would act as powerful extensions of physicians' intellect, assisting with diagnosis, treatment selection, and patient monitoring. The future seemed clear: medicine would become algorithmic, precise, machine-assisted.

By the late 1970s, two main approaches to automated diagnosis had taken shape. Both came up short. The first path led straight into "the mind of the expert." Rule after rule, each small and sharp-edged, strung together to form a lattice of medical logic. These systems, aptly named expert systems, tried to capture a physician's thinking as a chain of if-then steps.

*If the patient has fever, cough, and chest pain, then consider pneumonia.* A thousand rules like this, drawn from the neat lines of textbooks and the dense columns of journals, promised to turn a computer into a tireless diagnostician. The project seemed endless, but not impossible. It was just a matter of encoding enough expertise, one rule at a time.

Early systems did well on tidy tests. In narrow domains, such as selecting antibiotics based on symptoms and lab values, they sometimes matched specialists. The trouble came outside the test cases.

Real patients did not arrive in textbook order. Symptoms overlapped. Diseases hid. Rare conditions wore the costumes of common ones. To cover one specialty, the number of rules climbed into the thousands. Add a rule to handle a new case, and ten older rules contradicted it. The systems were brittle. Worse, they had no idea *why* their rules worked. They matched without understanding.

## THE PATTERN-RECOGNITION ALTERNATIVE

The second path tried a different tack. Skip the rules. Show the machine many examples of pneumonia, and let it spot the resemblance when a new patient walks in.

The psychologist Frank Rosenblatt built such a machine in the late 1950s and called it the perceptron. The design borrowed loosely from the brain. Each artificial neuron took several inputs, weighted them, added them up, and fired if the total cleared a threshold. Wrong guesses adjusted the weights. Right guesses left them alone. The machine *learned.*

The Navy paid for this work, hoping perceptrons might one day pick objects out of photographs. When Rosenblatt showed his Mark I sorting triangles from squares after a few dozen examples, the press lost its head. The New York Times reported that the Navy expected the new machine to "walk, talk, see, write, reproduce itself and be conscious of its existence."

The hype was decades early. A single-layer perceptron could only separate patterns that fell cleanly on opposite sides of a straight line. Anything more tangled, including questions as basic as whether two features both appeared, was beyond it. The pattern-matching machines had no grasp of underlying biology. They could note that fever, cough, and chest pain often arrived together. They could not see that bacteria were invading lung tissue, triggering inflammation, filling air sacs, casting shadows on a film. Without that causal picture, neither approach could reason about an unusual case. It also could not explain itself nor fill in the gaps when a patient's chart was incomplete.

## THE HARDWARE BOTTLENECK

Later systems added more biological knowledge. Performance improved, then hit a wall. The machines could not keep up with the math. Even simple reasoning about disease mechanisms required vast numbers of calculations. Testing a single hypothesis might require simulating dozens of biochemical processes, each with multiple steps. The computers of the day choked.

By 1987, Schwartz, so confident in 1970, conceded the obvious. Even on modern high-speed machines, the calculations were impractically slow. A system might grind for hours over a case a skilled physician could rough out in minutes. The hardware constraints, however, were not permanent. The evolution of computer storage and processing power followed a trajectory so predictable that it became a law of nature in the technology world.

In 1956, IBM introduced the Model 350 Disk File as part of the first commercial computer with a hard disk drive. This marvel of engineering, roughly the size of a large refrigerator and weighing over a ton, stored just under five megabytes of data. Five megabytes is barely enough for a few digital photographs today, or a couple of minutes of music. At the time, it was revolutionary. For the first time, computers could access stored information without rewinding magnetic tape, jumping directly to any piece of data in less than a second.

Progress was relentless. By 1980, IBM's 3380 disk drive stored one gigabyte—200 times more than the Model 350. The 3380 still weighed 250 kilograms, occupied several cubic feet, and cost $100,000. A single gigabyte, an amount of storage that today fits on a thumbnail-sized chip costing pennies, required an investment equivalent to a house and a reinforced floor to support it.

The force driving this transformation had been identified years earlier by Gordon Moore, co-founder of Intel. He predicted that the number of transistors that could be packed onto a chip would roughly double every two years. It was not a law of physics. It was an observation about money, chemistry, and clever engineering. As techniques improved, circuits could be etched finer. Finer circuits meant more transistors. More transistors meant more memory, faster processors, lower costs.

The prediction held for nearly half a century. Each doubling rippled outward. Speeds climbed. Memory ballooned. Power per calculation dropped. By the 1990s, a desktop machine outran the supercomputers of the 1980s. By the 2000s, a phone in a back pocket carried more computing power than the room of machines that had landed Apollo on the moon.

This compression of hardware didn't merely accelerate existing methods. It made entirely new methods possible. Algorithms that had been theoretical curiosities, too slow to run on real machines, became practical. Simulations that would have taken weeks finished in seconds. Data that would have overflowed all available storage could now be held and searched in memory. The strangling bottleneck of the 1980s dissolved.

## THE DATA DELUGE

For most of computing's life, data had been tidy. Library catalogs. Medical records. Bank ledgers. Scientific measurements. People typed it in, machines crunched it back out.

Then the internet opened, and the world began to overflow. By the 2000s, data was no longer something humans carefully entered. It was something that streamed out in billions of posts and photographs, a ceaseless outpouring of comments, emails, songs, sensor readings, and search queries. Digital cameras gave us the ability to pour images into vast, humming servers. Search engines became insatiable archivists.

Medicine joined the flood. Wrist-worn sensors started tracking heart rate, sleep, steps, and movement around the clock. Genome sequencing, once a decade-long project that cost a billion dollars, became a routine test that produced gigabytes per patient. Imaging moved from grainy X-rays to high-resolution CT, MRI, and PET scans, each producing hundreds or thousands of slices. Electronic health records captured diagnoses and prescriptions as well as vital signs, lab values, medication histories, and the free-form prose of clinical notes.

The constraint flipped. For most of history, the scholar's complaint was scarcity. Knowledge was rare, scattered across rare books and individual memory. Now the complaint was abundance. *How do you find a signal in a sea of noise? How do you spot patterns in a dataset no human can read in a lifetime? How do you train a system to recognize disease in millions of images that no team of radiologists can review?*

The answers would need more than faster chips and bigger disks.

They would require entirely new approaches to learning and intelligence.

## THE NEURAL NETWORK RENAISSANCE

In 1986, three researchers, David Rumelhart, Geoffrey Hinton, and Ronald Williams, published a method for training networks of artificial neurons arranged in layers. They called it backpropagation. The idea was to start at the network's output, see how wrong it was, and walk the error backward through the layers. Then nudge each connection by a small amount in the direction that would have helped. Repeat across millions of examples. The network slowly tunes itself.

In principle, a network with enough layers could learn almost any pattern. However, "in principle" and "in practice" remained far apart. The hardware of the time could not train large networks on real-world problems. Neural networks performed well on toy tasks, such as reading handwritten zip codes. They lost to other methods on almost everything else.

The breakthrough came in 2012, when a team led by Geoffrey Hinton unleashed a deep neural network, AlexNet, on the annual ImageNet Challenge. ImageNet was a vast digital menagerie, teeming with more than a million labeled images sorted into a thousand wildly varied categories. Dogs mid-leap, cats curled in sunbeams, cars gleaming under neon, mushrooms sprouting from damp earth, mosque domes catching the light, faces caught in every expression. The task was deceptively simple. Show the machine a photo, and ask it to name what it saw.

For years, winners had relied on hand-crafted tricks—edge detectors, texture filters, and clever engineering stitched together by human intuition. AlexNet ran them over. Its error rate of 15.3% beat the next-best entry by more than ten percentage points. Given enough data and enough compute, a neural network had just outperformed decades of careful engineering.

Three ingredients made the moment possible. The data: a labeled dataset of unprecedented scale. The hardware: graphics chips originally built to render video games, repurposed because their parallel design happened to match the math of neural network training. The training tricks: small refinements in how the layers were initialized and regularized.

The field reorganized itself around the result. Companies that had ignored neural networks rushed to hire deep learning researchers. Google, Facebook and Microsoft pivoted. Money poured in. By 2015, deep learning was the dominant approach in machine learning.

## THE ARCHITECTURE THAT CHANGED LANGUAGE

For several years after AlexNet, neural networks continued to excel primarily at perception tasks, such as image recognition and audio processing. Language remained more difficult.

Earlier text networks, called recurrent neural networks, read words one after another and tried to remember what came before. It's a bit like reading a book and writing notes on a magic notepad—one where each new note makes the last one fade away. By the time you finish a long chapter, you might only remember the last few lines. The beginning has already slipped from your mind.

A key change came when engineers moved from programming explicit rules to building systems that learn from data. This was a shift in philosophy. Stop telling the machine what words mean. Show it what words *do*.

A neural network trained on a vast pile of text does not start with definitions. It starts with numbers. Each word becomes a long list of coordinates, called an embedding. One modern model uses more than 4000 numbers per word. On their own, the numbers are noise. Together, they place words in a kind of mathematical landscape.

Think of this in terms of recipes instead of words. Imagine every dish from a huge cookbook as a dot on a big piece of paper. At first, the dots are scattered all over. Then, the system starts reading millions of menus and recipes. It pays attention to which dishes are served together in the same meal. Whenever two dishes appear together, the system gently moves their dots closer together. Over time, dishes that are often paired—like Caesar salad and caprese—end up side by side. This isn't because someone explained they're both salads. Instead, it's because they keep showing up together. In this way, the system figures out what dishes are related just by noticing their patterns.

Replace dishes with words and meals with sentences, and the same logic builds a map of language. *Dog* and *cat* drift toward each other. They both move away from *chair*. The coordinates of *dog* are not anchored to anything in the real world. They only describe how a *dog* sits relative to every other word the system has ever seen.

Once the words have positions, the model answers one question over and over: *given the words so far, what word is most likely next?*

The geometry of the map carries surprising structure. The vector for *king*, minus the vector for *man*, plus the vector for *woman*, lands near *queen*. The same trick works in our recipe space. A *burrito* minus the *tortilla* lands near a *burrito bowl*. None of this was programmed. It emerged from billions of examples of words and dishes in their natural company.

At its heart, a large language model is a next-word machine. Type a prompt, and it picks the most likely next word, then the next, then the next, building a reply one token at a time. Human language carries an enormous amount of what humans know, so the next-word machine often sounds as if it understands what it is saying.

On university exams and medical licensing tests, the best of these models can match or beat the average human test-taker. The leap in skill came from a particular design called the transformer.

The transformer threw out the read-one-word-at-a-time habit of earlier networks. In its place, it used a mechanism called self-attention. Every word in a passage could be compared directly to every other word, all at once. If a pronoun in sentence twelve referred back to a name in sentence three, the network could look straight at sentence three, no need to pass a memory through nine sentences in between.

This change let these models learn much faster on modern computers. It also meant they could read longer passages without forgetting what came before. Early transformer models raced ahead, quickly outpacing older systems.

Within a year, the architecture was pointed at a new target: huge models trained on the entire readable internet. In October 2018, Google introduced BERT, short for Bidirectional Encoder Representations from Transformers. BERT learned to read words in context by training on Wikipedia and a library of books. It became the engine behind Google Search's language understanding.

That same year, OpenAI released GPT, which stands for Generative Pre-trained Transformer. This was an early language model. The next version, GPT-2, came out in 2019 and was much larger. It used 1.5 billion settings, or "parameters," to make its guesses and could write surprisingly smooth and realistic paragraphs. By 2020, GPT-3 arrived, using a massive 175 billion parameters. It could do things it had never specifically practiced before, just by seeing a few examples. In other words, if you showed it a couple of math problems or writing prompts, it could pick up the pattern and try to follow along.

Bigger models, trained on more text, kept getting better. The hardware curve Moore had drawn in 1965, combined with the data flood unleashed by the internet, finally made it possible to build language models at a scale that produced startlingly good results.

## AN ARCHITECTURE OF FAILURE

Let's return to that plate we flipped at the start—the one with its tomato, carrot, and potato tumbling through the air. After all the history and breakthroughs, the most advanced language models still can't truly picture those vegetables smacking onto the kitchen floor. They can juggle words, predict sentences, and echo facts. However, when it comes to that simple, everyday moment, they're left staring at the mess.

Ask a language model to count only the vegetables that remain on the plate, and watch how its attention flickers. It latches onto certain words: *Vegetables. Remain. Top.*

Toss in a hint about botanical accuracy, and suddenly fruit jumps out—tomato is a fruit, right? The model, trained on countless articles and arguments about tomatoes, gets pulled into a trivia contest about labels. It's now missing the point entirely. Instead of picturing the tomato sliding off the plate and splattering on the floor, it's trapped in a debate about fruit versus vegetable.

The machine can't actually see what happens in the kitchen. It can't feel the drop, hear the splat or thud, nor can it weigh gravity against the rules of botany. It lives in a world made only of words. It shuffles them according to the patterns it's learned, never truly grasping the messy reality beneath. It has no sense of physical cause and effect. It only measures how each word affects the next prediction.

This is the pattern behind what is called *hallucination*. When word embeddings click together in a way that matches what a human reader expects, the output looks intelligent. When they do not, the output is confidently wrong. To the model, these two cases are identical. A hallucination is not a random malfunction. It is a maximally plausible prediction that happens to be false. The model is doing exactly what it was built to do

The benchmark called Simple Bench, which probes this family of failures, finds three big domains where the architecture cracks.

**Spatial reasoning** breaks when the model tries to track an object through real space. The plate of vegetables. A glass tipping at the edge of a table. A box sliding down a ramp.

**Temporal reasoning** breaks when the question depends on a sense of how long things take. Cooking a roast. Walking a mile.

Driving from one city to another. Text rarely spells out durations with much precision, and so the model fills the gap with plausible numbers it cannot check.

**Social-situational inference** breaks when the answer depends on simulating a particular person with particular knowledge in a particular scene. The model carries a great deal of sociological information in its embeddings. It cannot run the simulation that pulls that information into a specific moment.

Simple Bench keeps these three separate from shallower quirks. Models miscounting the letters in a word, for example, or insisting that 9.11 is larger than 9.9 because the string "9.11" shows up in training data tied to a major event. Those are different problems entirely.

These are not mistakes that one more round of training will fix. They reveal what language models are and what they are not. The same machine that can pass a medical board exam can fail at a question a five-year-old solves with a glance. Both results come from the same engine. Pattern matching across what the model has read.

Exam questions live in the training data, written down many times by many teachers. Everyday physical reasoning, like flipping a plate, is rarely described in text, so the model has nothing to draw on. Temporal failures follow the same shape. Language carries little exact information about durations. Phrases like '*took a few minutes*' or '*finished quickly*' map to broad, fuzzy ranges. Ask whether a specific task could be finished in a specific time, and the model will often invent a plausible answer that no part of its training can check. Social-situational failures arise when the question requires running a small mental model of a specific person. The architecture does not run mental models. It runs probability across tokens.

## THE TEST CASES

Medicine makes these strengths and limits concrete. From the early 2000s, machine learning crept into clinical practice through tasks that were narrow and repeatable. Reading electrocardiograms. Scanning retinal photographs. Spotting suspicious skin lesions. Counting white cells. The performance is not always perfect. It requires skilled oversight. It is good enough to matter in settings where rapid image reading was needed, and a local specialist was not available.

Digital pathology took things to another level. Not long ago, a pathologist would sit at a microscope, carefully peering through one glass slide at a time, hunting for signs of disease. Now, powerful software can race through thousands of tissue images in minutes, instantly flagging anything unusual and grouping similar cases together.

One system, called PathChat, combines a vision model trained on over 100 million images of stained tissue with a language model fine-tuned on more than 450,000 clinical instructions. When PathChat receives additional details—such as the patient's age, sex, or medical history—its accuracy on multiple-choice diagnostic questions jumps from 78% to 89%. What's remarkable is that simply adding everyday info, not fancy new image tricks, made the system much smarter at spotting what matters.

A model called life2vec stretches the language-model idea in a different direction. Trained on national registry records for millions of Danish citizens, it treats life events, employment changes, income shifts, hospital visits, and diagnoses as the equivalent of sentences. After training on records from 2008 to 2016, life2vec predicted with about 80% accuracy which people in the dataset had died by 2020. It identified factors linked to early death, including low income, male sex, and mental-health diagnoses. It also predicted personality traits, including extroversion. It missed clusters around accidents and sudden cardiac events, exactly the deaths that do not announce themselves in a sequence of paperwork.

On 207 multiple-choice questions drawn from *the New England Journal of Medicine* Image Challenge, GPT-4 with Vision matched physician-level accuracy at 82%, compared with a physician baseline of 78%. It got the answer right in 78% of the cases where physicians got it wrong. In 36% of cases where the model picked the correct final answer, its reasoning contained errors, with mistakes in image comprehension as the most common slip. A correct answer built on wrong reasoning is not understanding. It is a well-calibrated coincidence.

In another study, a chatbot built on GPT-4 was tasked with arguing against conspiracy theories. After a single 8-minute conversation, participants' confidence in their chosen conspiracy dropped by an average of 21%. A quarter of them moved from firm belief to genuine doubt. A follow-up survey two months later found the shift had stuck.

The system did not know any single conspiracy theory better than a well-prepared human counselor. What it had was patience and depth. It could deploy detailed counterevidence across thousands of conversations without ever getting tired or losing its place. The lesson was simple. Access to well-organized counterevidence mattered more than the perceived authority or warmth of the messenger.

## BARRIERS & OPPORTUNITIES

The cost of running these systems is also no small thing. Today's large language models live in data centers that glow and whir like a city that never sleeps. These machines gulp down thousands of gallons of water to stay cool and devour electricity at a stunning pace. Training a single large model can use as much electricity as hundreds of households burn through in a year. And the job isn't done once the model is built; every question we ask, every response it gives, adds to the bill. Multiply that by billions of conversations, and the numbers get staggering. This pressure is one reason researchers are working on smaller, more efficient designs. The goal is models that can do more with less.

Even as these models grow smarter, another truth comes into focus. Each new leap delivers smaller returns. The easy wins are gone. The best data is already in use. The costs—energy, computing, and human ingenuity—are rising. The road to more capable artificial intelligence may not be a straight path paved with bigger models and more data. There are bends ahead, and shortcuts are unlikely to get us there.

*So, where does that leave us?*

Today's language models have shown us the magic of patterns: they can translate languages, summarize pages of information, and even help diagnose disease. Yet, ask them to follow a tomato as it slides off a plate, and they're lost. That stubborn gap—between words and the world, between clever imitation and genuine understanding—is the frontier. The next great leap may belong to those who can teach machines to see and reason about the world we live in.

✧✧✧

# CHAPTER 2
# IS AI INTELLIGENT?

A SPECIFIC QUESTION comes up early in every serious conversation about artificial intelligence.

*Is artificial or machine intelligence* ***actually*** *intelligent?*

In ordinary speech, *intelligence* is a bag of abilities. Remembering. Inferring. Learning. Adapting. Planning. Imagining. The pieces tumble together, indistinct, refusing to sort themselves into tidy rows. No clean line separates one from another. Ask whether a computer system is intelligent, and the answer depends almost entirely on which ability the person asking has in mind.

Intelligence is quarried, chipped, and stacked by evolution over epochs. One ability lay atop another, like bricks in a wall rising from the mud. Each layer supports the next, a living architecture built by time itself. Understanding what that sequence produced, and what it required, helps clarify what we can expect from systems that arise by entirely different means.

## HOW INTELLIGENCE GREW

Biological intelligence, as biologist Max Bennett explains it, has five layers. Each is laid down on the one before it. First is steering.

**Steering** came into existence roughly 550 million years ago, a quiet revolution beneath ancient seas. For the first time, early animals weighed whether to move toward a flicker of light or slip away from a looming shadow, judging each stimulus as safety or threat. The ability seems humble. Yet, in that moment, life began to sketch its first crude maps of the world. Next came reinforcement.

**Reinforcement** is the learning-by-consequence that stitches experience to action. Creatures learned: pain signaled retreat, food called them back. It was like a child wobbling on a bicycle—tip, fall, recover, try again. The world became a classroom where every stumble or success etched another lesson. This trial-and-error loop laid the foundation for all the cognition that would follow. Then, simulation dawned.

**Simulation** is where creatures gain the power to run silent rehearsals in the mind, previewing danger or reward before their bodies leap. Every decision could be staged in imagination first. It's a private dress rehearsal, where missteps cost nothing, and catastrophe could be sidestepped before it ever arrives.

**Mentalizing**—what psychologists now call theory of mind—let animals look behind the eyes of others, modeling not just bodies but minds. Fellow creatures were no longer just obstacles or scenery. They became agents. They are beings with their own hungers, plans, and secrets, capable of surprise and strategy. Last came language, the most astonishing leap.

With language, suddenly, knowledge could leap from mind to mind, untethered from the slow cost of experience. One animal could warn another of a lurking predator, offering a lifeline won by one and shared by many. Survival knowledge was unearned, yet freely given. With language came the miracle of cumulative culture.

Knowledge does not just survive. It grows with every generation. No longer did each lineage start from scratch. The first stone tools required just two or four steps to make. Fast-forward to 600,000 years ago, and tools demanded a dance of more than ten precise moves—complexity beyond even the cleverest non-human apes. By

300,000 years ago, our ancestors' inventions had left all rivals behind. Each leap in toolmaking was knowledge rescued from oblivion, preserved and handed forward, so that what one mind discovered, many could inherit and extend.

This long evolutionary ascent shows that biological intelligence is not a single threshold. It is a scaffold built slowly over time. With this foundation in mind, the questions grow sharper as we turn from the past to the present. *What, exactly, are we pointing to when we talk about intelligence? What does it mean for a mind—biological or machine—to be aware?*

## AWARENESS & CONSCIOUSNESS

Are you aware right now? *Yes.*

Are you conscious? *Also yes.*

Are these the same answer to the same question?

The words slide past each other in everyday use. *Awareness. Self-awareness. Consciousness.* We use them interchangeably, as if they pointed at the same thing. They do not. The distinctions among them are subtle, and they matter—for neuroscience, for clinical medicine, for the philosophy of mind, and for the question of what, if anything, a machine system might one day possess.

**Awareness** is the registration of something. A frog is aware of a fly. A self-driving car is aware of the lane markings. A thermostat is aware, in a stretched sense, of the temperature. To be aware is to have a state that tracks something else. The thing being tracked may be external, such as a sound, a face or a hazard. It may be internal, like a hunger, a pain or a memory surfacing. Awareness, in this minimal sense, is widespread. It does not require a self.

**Self-awareness** is the registration of oneself as the one doing the registering. It is the mirror turned inward. A self-aware system does not merely perceive a fly; it perceives itself perceiving the fly. It can recognize its own body, its own thoughts, its own desires as belonging to it. The classical experimental test is the mirror test.

*Does the animal, on seeing its reflection, understand that the figure in the glass is itself?* Some apes pass. Some elephants pass. Some dolphins, magpies, and possibly cleaner wrasses pass. Most animals do not. Human infants begin to pass somewhere around eighteen to twenty-four months, around the time they start using their own names.

**Consciousness** is the broadest and most contested word. In one usage, it simply means being awake rather than asleep. "The patient is conscious", meaning responsive. In another, it means having any experience at all, in the qualia sense. This means there is something it is like to be the system in question. In a third, it means full subjective life, with self-awareness, narrative, memory, intention, and inner monologue all integrated. The same word does all three jobs, often in the same paragraph, and most disputes about consciousness are partly disputes about which job is meant.

Awareness is about something. Self-awareness is awareness of oneself. Consciousness is the broader field in which awareness, self-awareness, and experience occur. Three brief portraits make the distinctions concrete.

The frog is aware. The world enters its eyes; flies are detected, and tongues are launched. Threats are evaded. The animal navigates an environment with skill. There is no strong evidence that the frog is aware of *being the frog*. It does not appear to recognize itself in a mirror. It does not, so far as we can tell, model its own thoughts. Awareness without self-awareness.

The bat is aware in a way we cannot share, building a world from echolocation. It may or may not be self-aware. The relevant experiments are difficult to perform on bats. However, there is, almost certainly, *something it is like* to be a bat... some field of experience accompanying the awareness. Awareness, possibly self-awareness, is almost certainly some kind of consciousness in the experiential sense.

The person is something else entirely. Aware of the world, aware of the body, aware of thoughts, aware of b*eing aware*. Self-aware in the strongest sense. Capable of naming the self, narrating its history, modeling its future, recognizing its image, distinguishing its desires from those of others. Conscious in all three senses: awake, experiencing, integrated.

### Floor Plan: Awareness, Self-Awareness, Consciousness

Picture a building rising floor by floor from solid ground. Awareness is the foundation. It is the first stone laid, a broad floor spanning the landscape of all living things. Self-awareness is a higher story. It's added with care, allowing the structure to look inward, to recognize its own rooms and windows. Consciousness is the sweeping architecture that encloses every floor. Within this living edifice, awareness and self-awareness find their place, each level supporting and shaping the next, all held within the greater design.

It is possible to have lower floors without higher ones. Some rise only a floor or two above the ground. Awareness alone, with no higher stories. Others add a second level, peering into mirrors, glimpsing themselves in the glass.

It is not obviously possible to have higher floors without the lower ones. The upper stories cannot stand without what lies beneath. A tower of self-awareness without an underlying floor of awareness is a contradiction in blueprints. It cannot exist. Yet ground-floor structures—creatures, systems, minds—are everywhere, getting on with the business of living without ever climbing higher. *When considering this, where do today's AI systems fall?*

## A MENAGERIE OF MACHINE INTELLIGENCE

The animal kingdom is a motley of astonishing diversity—creatures that crawl, soar, swim, and sing. Each is perfectly tuned to its own world. Artificial intelligence, too, has begun to branch and flourish. It is spawning its own menagerie in the form of systems that see, reason, move, and imagine in ways as varied as the species of Earth.

Until now, our journey has wound through the worlds of computers, neural systems, and, finally, the architectures of large language models. The landscape of artificial intelligence stretches far wider. Beyond these are reasoning models, agentic systems, vision models, machine-learning engines, generative models, and robots that walk, grasp, and sense. Each is a distinct kind of system in this unfolding menagerie. Each is built from code and circuits rather than cells and nerves. Each has its own way of "being in the world" functionally.

A language model is a text-dwelling creature, tuned to read and write. A reasoning model is a planner that loops through steps and tools to pursue goals. Other systems—the seers, movers, and predictors—handle vision, action, and forecasts in their own specialized ways. This is similar to how a bat, a frog, and a person are all animals yet live and sense the world in very different ways.

## Large Language Models

A large language model like ChatGPT is only aware in a thin sense. Instead of truly understanding the world, it takes in words as pieces called tokens. It then finds patterns in what it has read and learned and predicts what should come next. It does not see or hear in the usual sense. Its "world" is the statistical shadow of language, not a lived environment.

Think of them as autocomplete powered by the whole internet. They do not understand the world. They cannot see when something does not fit. They cannot observe. They cannot run experiments. They predict the next word, again and again, based on what they have read.

That makes them sound knowledgeable, because they are looking backward at a lot of writing. It also keeps them from seeing forward. They cannot register a new problem in the real world because they have no way to sense one. They cannot form or test a guess about something that has not yet been described in the text. Their job, by design, is to keep predicting the next word in a sentence.

The distinction matters most at the frontier of knowledge. Where language describes a domain well, the models perform well. They summarize. They extend arguments. They restate complex material in clearer prose. Where language falls short, the model offers nothing. When an observation is not recorded, the model cannot register the anomaly or feel the puzzlement that would prompt a scientist to ask a new question.

Language models appear to be reasoning when they tackle a new problem. They produce plausible steps. They arrive at conclusions. The steps imitate patterns from training data. Some advanced models try to lean against this limit by being trained to show their reasoning. They break a big problem into smaller steps rather than leaping to a conclusion. It is the machine equivalent of writing out the work on a math test. It catches mistakes. It makes the process readable. Even here, the model is still predicting the next word and the next sentence.

Sometimes it may seem to understand the conversation, or to display *situational awareness*. In some instances, it even appears capable of metacognition, or of talking about its own limitations and strategies. However, these are best understood as pattern-completion behaviors on text, not as evidence of an inner subject that knows it is knowing.

As for self-awareness, large language models can describe themselves, their architectures, and their training in great detail. This knowledge is not bound to a persistent, felt self. It is just more text, pulled from a probability cloud. Philosophers call such entities "behavioral zombies." These systems pass many verbal tests for awareness and even self-awareness yet lack a plausible mechanism for subjective experience or integrated selfhood. Mainstream scientific positions converge, at least for now, on the view that current language models do not meet the necessary conditions for consciousness. This is because they lack continuous, integrated, world- and body-embedded processing that is stably tied to a single enduring agent.

If we place a large language model into our floor-plan

metaphor, it occupies something like a *simulacrum floor* beneath the frog. It reliably mimics discourse about awareness and self-awareness. It does not furnish independent evidence of either. It behaves as if there is "something it is like," but we have no theoretical or empirical grounds to think there is anything it is like to be the model.

## Reasoning Models

Reasoning-augmented systems, or reasoning models, are large language models scaffolded with tools for multi-step reasoning, external memory, code execution, or planning. Put more simply, if a large language model is a supercharged autocomplete, a reasoning model adds extra steps to that. As we saw in the previous chapter, language models struggle with problems that require multiple inference steps. Reasoning models break big problems into smaller parts, checks its own work, and use tools.

Take a small riddle, for example. *Julia has two sisters and one brother. How many sisters does her brother Martin have?*

Even the smartest version of GPT-4 is tripped up, stubbornly saying "two" over and over. However, the real answer is three—because Martin's sisters are Julia and the other two. Getting there means using some form of reasoning. Reasoning, however, isn't simply one thing. It's a whole toolkit, with different tools for different jobs.

**Deductive reasoning** is the most famous tool. Sherlock Holmes famously uses it in Sir Arthur Conan Doyle's stories. Deductive reasoning starts with a general rule like "*All the footprints are muddy, so the suspect must have come from the garden.*" It then uses it to reach a specific, logical answer.

**Inductive reasoning** works the opposite way. Instead of starting with a rule, you start with clues or examples. You then use them to spot a pattern or guess a rule. For instance, if you see the sun rise every morning, you might conclude it will rise again tomorrow.

Scientists use induction all the time, spotting trends in data and building general principles from repeated observations.

**Abductive reasoning** begins when something does not fit. Deduction cannot account for it. Induction has no precedent for it. So abduction steps in and makes the best available guess about what might be going on, even before proof is in hand. One example is Alexander Fleming spotting a halo where bacteria had died around a blue-green smudge of mold. He guessed the mold was releasing a toxin lethal to its neighbors. Thus, penicillin was discovered.

**Reductive reasoning**, by contrast, breaks things down into their simplest parts to understand how they work. It's like taking apart a clock to see how the gears move. This approach is common in science and engineering, where complex problems are solved by analyzing and reconstructing the basics.

Other, less well-known types of reasoning include analogical reasoning (solving new problems by comparing them to old ones), causal reasoning (figuring out what causes what), and common-sense reasoning (using everyday knowledge to make decisions).

Back to our earlier riddle about Julia and her family, we see the language model struggle because it cannot leverage any of our reasoning styles. To grasp the answers to these types of queries, the model must "hold" the whole family in mind and view it from Martin's perspective. This is reasoning from a shifted vantage point.

It is here that new reasoning models shine. Instead of just blurting out an answer, they break the puzzle down into steps. They generate what engineers call a chain of thought. For the problem, a reasoning model might say: *Julia has two sisters. Including Julia herself, that's three girls. Julia has one brother, Martin. From Martin's view, his sisters are the three girls. So Martin has three sisters.* The answer comes at the end of this careful, step-by-step consideration, not from just spotting a pattern.

Functionally, this looks more like human-like reasoning: decomposing tasks, checking results, iterating. From a distance, a reasoning model can mimic the frog's responsive skill and some of the person's narrative self-monitoring. Still, not everyone is convinced that this is true reasoning. Some see it as a clever imitation; machines mimicking the way people think out loud, without actually understanding what they're doing.

Take philosopher Shannon Vallor. She describes the chain-of-thought process as a kind of meta-mimicry. The systems, she argues, produce plausible-sounding reasoning traces that imitate the human habit of thinking aloud, without necessarily doing general problem-solving underneath.

Others argue these models are making real progress, beginning to reason in ways that matter. An OpenAI spokesperson noted that users build trust in a model's answer by watching how it *reasons through* its *response*. Trust, in this landscape, is often a matter of showmanship. We see a model lay out its logic, step by step, like a magician revealing a trick. Yet, the explanation it offers may be little more than a polished performance. Beneath the surface, the real machinery hums out of sight, and the story we are told does not always match the story being lived beneath the hood.

This gulf between appearance and reality—between the visible chain of thought and the hidden gears—leaves us squinting through frosted glass. This lack of transparency is a constant source of tension. Most leading models from companies like OpenAI, Google, and Anthropic keep their inner workings hidden. Outside researchers are left to study shadows on the wall, guessing at causes from effects. DeepSeek's R1 is a rare exception, inviting outsiders to peer directly into its mind. With that rare transparency, the gap between story and substance finally begins to close. Speculation can become understanding.

When compared to language models, the extra steps of a reasoning model serve performance, not experience. The system has richer meta-cognitive *behavior*—it can "talk about what it is doing" and even inspect its own steps. Under most current theories, consciousness requires more than structured access to internal states. It requires that these states are globally broadcast, integrated with embodied perception and action, and unified into a continuing subject. Today's reasoning models are closer to elaborate cognitive prostheses for humans than to standalone subjects. They extend *our* agency and awareness without convincingly instantiating their own.

## Other Models and Architectures

Other systems span a much wider range of quasi-awareness. Narrow perception models—vision classifiers, speech recognizers, motor controllers—resemble islands of frog-like awareness. They are tightly tuned to particular patterns, dynamically responsive, but devoid of evidence for a self-model or experience. Embodied agents that couple perception, memory, and action in robots or virtual environments approach an animal-like functional organization. Still, there is no consensus that such organization alone suffices for consciousness without biological substrates or specific kinds of information integration.

The hypothetical artificial general intelligence sits as the analog to the person. Artificial general intelligence, if it arrives, would in principle match or surpass human performance across nearly all cognitive tasks. It also exhibits flexible, cross-domain learning and planning.

Whether a machine could actually be conscious depends on which story about consciousness you believe. Functionalists say that if a machine's mind works in the right way—even if it's just code—it might be enough for real consciousness. Others argue that only brains built like ours, out of living cells or special brain-like chips, can ever truly feel or experience anything.

To fully explore this debate, we have to step back and ask: *what biological makes consciousness possible in the first place?*

Medical imaging paints a vivid picture: the self is not a single spot in the brain. It is a living pattern of activity that flickers and shifts as you move through your day.

Picture a hidden spotlight in the brain. Whenever you pause to think about who you are, that spotlight glows. Remember a moment from your past? It shines again, illuminating the memory as if replaying a scene on a stage. Imagine your future, step into someone else's shoes, or weigh a difficult moral choice—each time, this same hub gleams to life.

This shining hub is known as the precuneus, a region of the medial parietal cortex and one of the most active regions of the resting brain. When it's damaged or disconnected, people can lose their sense of self. They may struggle to recognize their own face, remember their past, or even feel that their body belongs to them. Whether in deep sleep, under anesthesia, or after a coma, the return of consciousness appears to go hand in hand with the reawakening of the precuneus and its surrounding network.

If one takes this pattern too literally, it tempts a strong conclusion: consciousness is whatever lights up when the precuneus is online. In this view, consciousness would be a local biological property, tied to a particular patch of cortex and the chemistry that keeps it firing. No precuneus, no consciousness. No human neural tissue, no genuine awareness of any kind.

Yet, the brain itself offers counterexamples.

At times, these parts of the brain can slip into a kind of pseudo-pause—like when we enter what's known as flow. Flow states tend to occur when several conditions are met. The task is challenging enough to require full attention yet not so difficult as to overwhelm. The goals are clear. The feedback is immediate. The actor's skills are matched to the demands of the activity. Under these conditions, something shifts. Self-awareness recedes. Performance often improves. The experience becomes deeply pleasurable.

It is not unique to elite performers. A teenager absorbed in a video game can be in a state of flow. A cook chopping vegetables in rhythm can be in flow. A reader lost in a novel, a programmer in a difficult debug, a child building with blocks. The state is widely available. It is also, in modern life, increasingly rare — the conditions that produce it are not the conditions that fill most days.

Flow is not a loss of consciousness. People in flow are not absent. They are intensely present. This illustrates that our consciousness is layered. It also shows, perhaps more clearly than any other ordinary state, that the experiential field can persist and even intensify while the reflective self-monitoring layer is downregulated. The taxonomy of awareness, self-awareness, and consciousness is not abstract. It is something the brain rearranges in real time as you climb a wall or sit down to write.

If consciousness simply *were* the activity of this one hub, then whenever its activity drops, so should "you." Patients whose precuneus is injured but gradually recovering would, on this strict view, pass in and out of existence as persons while their tissue heals. Musicians and athletes in deep flow—whose self-related networks quiet as performance peaks—would be considered less conscious precisely when their experience is most vivid. *Was Mozart unconscious in the middle of composing, simply because he was not narrating himself to himself?*

The intuition rebels. Our ordinary usage does too. We do not say that the climber in perfect flow is less conscious than the bored commuter refreshing a phone. These cases suggest a softer, more layered picture. The precuneus and its neighbors are not the entire story of consciousness. They are one crucial part of a broader system that coordinates self-related processing. Damage there can shatter the self-model without erasing every trace of experience. Flow can dim self-reflection while leaving raw awareness and rich experience fully lit.

For the philosophy of mind, this matters. Arguments about consciousness often equivocate among the three meanings. A claim that "machines are not conscious" may be a claim about machine awareness, about machine self-awareness, about machine experience, or about all three. The arguments differ depending on which meaning is in play. A machine that lacks experiential consciousness may nonetheless exhibit awareness and a form of self-awareness; a machine that has all the functional markers of self-modeling may nonetheless lack qualia. The categories must be separated before they can be debated.

Once we separate those layers, the strong biological claim—that consciousness exists *only* in brains with a functioning precuneus—begins to look less like a discovery and more like a stipulation. It may turn out that biology, as we know it, is one way to build systems with awareness, self-awareness, and experience stacked together. It does not automatically follow that no other architecture could ever realize some or all of those layers. When we later turn to the different kinds of AI systems like language models, reasoning models, and others, the question may not be "*Do they have a precuneus?*" but "*Which layers of this stack, if any, do their architectures actually support?*"

It is conceptually possible that, in the future, some artificial architectures will climb the floors of our earlier floor plan metaphor. They may ascend from basic awareness toward full, integrated selfhood. If such systems emerge, they may possess integrated, embodied, self-modeling capacities that support a unified subject. At this point, the conversation takes a turn. We move from the mysteries of awareness and consciousness to another central puzzle—intelligence itself.

## FROM CONSCIOUSNESS TO INTELLIGENCE

*What does it mean to call a system intelligent, as opposed to conscious?*

The distinction matters, because while the two concepts are often intertwined, they refer to different capacities. Consciousness is about having experience; intelligence is about the ability to adapt and solve problems. Acknowledging how these concepts relate, and where they diverge, sets the stage for the next phase of our inquiry.

To understand intelligence more fully, it helps to look beyond ourselves. Animals other than humans make it harder to treat human intelligence as the single reference point. They also give us a way to test a simple idea that general intelligence, whatever form it takes, rests on a small set of core capacities. For our purposes, we will focus on three that appear again and again in the literature and in the lab: the ability to generalize across situations, to represent the world in internal models, and to select what matters from a flood of information. Rather than define these pillars in the abstract, we can watch them at work—in beaks, fins, and arms.

Corvids, the famously clever clan of crows and ravens, naturally operate like tacticians. They scatter food across hundreds of hidden caches, each one tucked away like a secret they refuse to confess. Months later, they return with eerie precision, as if guided by an internal map etched in memory.

Their intelligence doesn't stop at recall. If another bird lingers nearby, watching with too much interest, the corvid adapts. It delays the cache, relocates it, or feigns indifference until the coast is clear. This is generalization in feathered form.

**Generalization** is what lets intelligence jump the rails of experience. A crow that learns to hide food in one kind of landscape and then applies the same caching tricks in a new one is doing what a child does when they move from sorting fruit to sorting tools: pulling a pattern out of one setting and carrying it into another. It is the ability to see the rule behind the situation and to apply it somewhere new.

Yet intelligence is not confined to the skies. Beneath the waves, a different kind of mind emerges. Dolphins form intricate social networks that rival those of many land animals. Each individual develops a signature whistle, a kind of name learned early in life and recognized for decades. Their knowledge is passed from one to another in a fluid exchange, like lessons shared in a boundless, shifting classroom. A dolphin coordinating a group hunt must track each partner's location, where the fish are likely to move, and how its own motion will change the scene. That requires an internal map of bodies and currents, a simulation of possibilities. This is known as representation.

**Representation** is the engine behind anticipation and planning. It's the quiet "what if?" that runs in the background before any leap or lunge. It is also the mind's knack for building inner worlds.

Dolphins don't hold a monopoly on the intelligent minds of the sea. Lurking in rocky crevices and swirling currents, octopuses reign as the escape artists of the deep. They are loners by nature. Alien to us in both form and mind, their lineage split from ours 750 million years ago. Yet their arms can curl and uncurl with purpose: opening jars, solving mazes in a single try, or playing with found objects just for the joy of it. Their intelligence flashes out in moments of curiosity and mischief. In them, we see selection.

**Selection** is the art of knowing what matters and what to ignore. An octopus emerging from its den faces a sea of signals: shifting light, drifting sand, the movements of prey and predator. It does not respond to everything. It chooses. It attends to the promising shell, the gap in the rocks, the glint of a crab. Without this selective focus, intelligence would drown in detail.

Every one of these creatures incorporates the three pillars—generalization, representation, and selection—into its own unique pattern. Their abilities are shaped by the worlds they inhabit, not by any blueprint borrowed from humanity. Intelligence, it turns out, wears many faces, and it doesn't need to look like ours to be real.

A crow and a mathematician are both intelligent. Their genius simply shows up in different arenas. One in the rustle of feathers and clever puzzles, the other in numbers and proofs. Machine intelligence is another step along this branching path. A language model won't soar through the sky. Instead, it can answer questions and organize knowledge on a scale no human ever could.

Today's machines can approximate parts of these pillars. They can generalize within their training data. They can build fragments of inner models and make choices that look selective. Yet when the ground shifts too far from what they've seen, the scaffolding wobbles. No machine, yet, stands tall on all three pillars in the way animals routinely do. To see exactly where they stand, we have to move from forests and oceans to test rooms and screens—from lived intelligence to measured performance. That is where benchmarks enter the story.

## BENCHMARKS & LIMITS

Benchmarks are standardized tests that aim to measure what a system can actually do, beyond the electricity it consumes or the hardware it runs on. They give us a way to compare different models on shared challenges, and to see how far machines have come and how far they still have to go. One example of these benchmarks is progress on composite cognitive tests.

One measure of machine intelligence is through the AGI score. AGI stands for "artificial general intelligence". The AGI score assesses 10 mental skills—such as reasoning, language, and planning —and tests whether a system can reach the level expected of an average human adult. For example, the popular AI model GPT-4 scores about 27% on this test, while the next version, GPT-5, scores around 57%.

For comparison, these systems currently automate about 2.5% of freelance work, such as online writing and editing jobs. Still, there's a big gap between getting a good score on a test and actually doing well out in the real world. It's like getting a high score on a practice driving test, and still struggling when you're on the road in busy traffic. Closing that gap will take more than "good" test results. A later chapter explores this challenge in depth.

It's also important to realize that not all tests are created equal. Take ARC-AGI, which stands for "Abstraction and Reasoning Corpus – Artificial General Intelligence." This special benchmark is designed to test something called fluid intelligence. Fluid intelligence is the ability to look at a few examples, spot the pattern, and figure out the rule, even if you've never seen anything like it before. The puzzles in ARC-AGI are tricky because they don't use regular words or familiar shapes. You can't simply guess based on past training. Most humans can finish most of these puzzles. Top language models, despite doing great on school exams and coding challenges, barely do better than random guessing on ARC-AGI.

The latest version, ARC-AGI-3, released in March 2026, makes things even tougher. Instead of static puzzles, the test now puts models in interactive environments. The system has to make a move, see what happens, and then adjust its plan on the fly—much like playing a video game, where you learn the rules as you go. So far, even the newest models still lag far behind humans in these situations. The gap shows that while machines are getting smarter, there's still a long way to go before they can match the flexible, creative thinking people use every day.

Forecasts now suggest a 10% probability of reaching general machine intelligence by the end of 2026, a 50% probability around 2041, and a 90% probability by the 2160s. The figures are compressed from earlier projections centered on 2060. However, these predictions are still less optimistic than those of some companies, which sometimes say that general intelligence is already here or will arrive very soon.

This benchmark cycle carries its own lesson. Scores rise faster than real capability when labs optimize for the specific test rather than for the underlying cognitive property the test was built to measure. The dynamic, known in economics as Goodhart's law, shows up reliably in machine-learning evaluation. A score that signals strong general thinking skills stops meaning the same thing once people know the model is trained specifically to do well on that test.

The useful question is not which system leads the current leaderboard. The useful questions are what assumptions, resource limits, and scaffolding choices the team needed to take the lead, and how closely that process resembles what humans do when they solve the same problems. Sometimes, we need a different kind of benchmark altogether.

## A Benchmark of a Different Kind

Zen Buddhism uses koans. These are questions that are not designed to have answers. They are designed to train the mind in the practice of sitting with uncertainty.

The 18th-century Zen master Hakuin Ekaku asked his students, "*What is the sound of one hand clapping?*"

The question is a riddle made of smoke—deliberately unanswerable by ordinary logic. The student who approaches it with deduction finds the trail vanishing beneath their feet, no rule to grasp. The one who tries induction reaches for a pattern and closes their hand on emptiness.

Pose the koan to a large language model, and a different kind of theater unfolds. One system offered several candidate responses, such as silence, representing the absence of a second hand. It also offered a symbolic gesture of one hand extended in the air.

The system also identified the philosophical concepts the koan is built to evoke. Non-duality. The challenge to binary thinking. Interconnectedness. The dependence of phenomena on relational context. The limits of logic in understanding certain aspects of experience. The invitation to move from intellectual analysis to direct encounter with the world.

The response in this case drew on extensive training data about Zen Buddhism, the history of koans, the psychological literature on insight, and the philosophy of paradox. It was also not what a human student sitting with the koan for months would experience. The model retrieves and presents. It does not sit with the question. It does not encounter the frustration that, according to the tradition, is the beginning of the insight the koan is meant to catalyze.

Neuroscience offers a window into how those magical flashes of insight happen in the brain. When someone suddenly "gets it"—not by working through every step, but in a single leap. It's as if a lightning bolt fires across the brain. Special brainwaves, called gamma-band oscillations, ripple through the cortex. Blood rushes to the right side of the brain, lighting up the area that helps us see new connections. Unlike ordinary logic, which builds answers piece by piece, insight is like connecting dots in a single swoop. That "aha!" moment is your brain making a fast, creative guess, skipping the slow, step-by-step path.

Current language models have no equivalent. The architecture processes tokens in sequence. The attention mechanism weights relationships between positions in a sequence. There is no mechanism that resembles the broad integrative activation that characterizes human insight. What the models produce, when given a koan, is the most probable continuation of the text pattern the question opens. The response looks like an answer. Whether it reflects anything resembling insight is a question the output cannot settle.

The observation does not erase the practical value of the model's output. A student exploring the koan for the first time finds the model's map of candidate interpretations genuinely useful as a starting point. The model offers the intellectual context that a student needs before beginning the meditation practice. What it cannot offer is the practice itself, or the understanding that, in the Zen tradition, the practice is meant to produce. There is a difference between knowing about a territory and walking through it. That difference is exactly what separates current machine systems from the full cluster of abilities the word *intelligence* tries to name.

These benchmarks and tests shine a light on what today's systems can and can't do, revealing both their growing strengths and stubborn gaps. Yet, the story of intelligence isn't measured by abilities alone—it also hinges on the resources that bring those abilities to life. To see the full picture, we need to move from what systems achieve to what it costs them to achieve it.

## The Watt Problem

The human brain runs on roughly 20 watts. About the same draw as a dim incandescent bulb. On that budget, it handles perception, motor control, emotional regulation, memory consolidation, and all the cognitive work humans accomplish in a day.

Current frontier machine models live in data centers that draw megawatts per training run and consume significant power per query, depending on the model and the hardware. Training GPT-3 alone consumed an estimated 1,287 megawatt-hours and produced about 502 metric tons of carbon dioxide. Individual inference accelerators draw 700 to 1,200 watts each. A single query against a frontier model passes through tens or hundreds of them in parallel. Industry projections place global demand for machine-learning power at roughly 68 gigawatts by 2030.

The most capable machine systems in deployment today consume energy at a rate several orders of magnitude above the brain's. Measured in terms of cognition per watt, current machines are not in the brain's league.

The gap is not an economic inconvenience. It is a signal about architecture. The brain achieves its capacities through sparse activation, analog computation, and parallelism, all shaped over millions of years of metabolic pressure. A system that reproduces some intelligent behaviors while burning millions of times the energy is not a larger brain. It is a different kind of mechanism, optimized along different axes.

Whether energy efficiency should count as a criterion of intelligence is a definitional question. If it does, the gap between current machine systems and biological intelligence is wider than benchmark scores suggest. If it does not, the gap is narrower. As we look ahead, the conversation shifts from what today's machines achieve to what might be possible in the future. The idea of superintelligent systems —machines that could far surpass human abilities in almost every area—raises new questions and possibilities.

## THE SUPERINTELLIGENCE HORIZON

In 1965, the British mathematician I.J. Good described the possibility of an intelligence explosion. A machine slightly smarter than a human redesigns itself, improves its own algorithms, and enters a positive feedback loop of rising capability. *The first ultra-intelligent machine,* Good wrote, *is the last invention that man needs ever make.* The recursive self-improvement he described would compress what might otherwise take centuries into hours or days.

Five dimensions help bring the idea into focus.

**Speed.** Silicon circuits switch about a million times faster than biological neurons fire. At that ratio, a machine on suitable hardware processes a year's worth of information in roughly 30 seconds of real time.

**Generality.** Where any given human specializes, a sufficiently capable system applies its full cognitive resources to any domain at once.

**Self-improvement.** Biological brains change on evolutionary timescales. A digital system rewrites its own code and designs better hardware for the next version of itself.

**Conceptual reach.** Higher intelligence reaches problems that lower intelligence cannot. The way quantum mechanics and eleven-dimensional string theory sit beyond the reach of most humans.

**Time horizon.** Unbound by mortality, a system can hold a problem in mind across centuries, a resource no individual human has.

Each dimension describes a different relationship to knowledge and time than we currently know. Shakespeare was a literary genius with no special mathematical gift. Einstein was a physicist, not a composer. A sufficiently capable system brings full cognitive resources to every domain at once, without the specialization that an individual human lifespan forces. Whether that counts as intelligence in any meaningful sense, or as something for which human language has no good name yet, is a question current frameworks cannot answer.

## THE HIDDEN SYSTEMS AMONG US

We began this chapter asking what it means for a machine to be intelligent. Along the way, we discovered that intelligence is a sprawling landscape. It's a set of abilities built on layers of awareness, memory, reasoning, and adaptation, each shaped by their own history and environment. No machine yet walks that landscape as fully as a human or a crow, but the boundaries are shifting.

Yet as we debate tests and definitions, a more urgent question rises: *What does it mean to share our daily world with these systems, even if they do not mirror us?*

Artificial intelligence has slipped quietly into so many corners of modern life that its presence is both obvious and almost invisible. It is embedded in the rhythms of our routines, humming beneath the surface of our choices and interactions.

Most of us already live with forms of "hidden AI." We navigate traffic guided by algorithms, see our words corrected as we type, accept or reject recommendations for music, news, and even friendships. All of this is shaped by machine systems we rarely notice.

The intelligence on display does not gather itself into a single mind or body. Instead, it flickers through networks and devices, sometimes acting on our behalf, sometimes steering us in directions we do not see. In this sense, artificial intelligence is not just a future possibility—it is already part of the fabric of our world, present yet hidden in plain sight. In the next chapter, we turn to these hidden companions.

✧✧✧

## CHAPTER 3

# THE HIDDEN AI ALREADY AROUND YOU

IN THE SUMMER OF 2024, Apple quietly reshaped the digital landscape. The company announced that its next software release would usher generative machine intelligence into iPhones, iPads, and Macs. It embedded a new capability called Apple Intelligence. This was not a distant promise, but an imminent arrival.

Most technological changes announce themselves with spectacle and sound. The locomotive arrived trailing a plume of smoke and the shriek of steel. The automobile rolled down city boulevards, drawing crowds. The personal computer commandeered a place on the desk, humming with promise and possibility. This change was different.

It slipped into homes invisibly, crawling across the world's devices at three in the morning on a Tuesday, while phones recharged on bedside tables. By the time the sun rose, every email's autocomplete had been subtly, silently altered by a system trained on the collective record of human expression. The only clue was a slight improvement.

The system could now write emails, transcribe phone calls, and sift through the endless notifications piling up on lock screens. With billions of Apple devices scattered across the globe, the news signaled more than a product update. It was a tectonic shift. Generative machine systems, once something users had to seek out, would now slip quietly into daily life, woven seamlessly into the hardware people already carried.

This was the hidden moment of machine intelligence. The technology seeped beneath the surface, dissolving into the infrastructure below everyday awareness. The story, however, does not begin with iPhones. To understand what makes this change so significant, it helps to step back and ask: what does it mean for any species to store and transmit knowledge? Before we can appreciate how machine intelligence rewrites the rules, we need to see how the rules were first written—starting long before humans, in the world of animal minds.

## THE LONG ACCUMULATION OF KNOWLEDGE

Pigs root through earth with sticks clamped in their jaws. Bumblebees, after watching a clever peer, tug strings to win a drop of sugar water. Sperm whales sing in dialects that drift across ocean basins—one pod echoing the marine equivalent of a Boston accent, another a Glasgow drawl. Yet these flashes of insight scatter and fade.

What sets humans apart is not the tools themselves or the learning. It's the relentless layering of knowledge. Each generation stacks its discoveries atop the last. Culture, for our species, is a living archive. Wisdom is refined, recorded, and bequeathed, so that no child needs to invent fire or language anew.

A chimpanzee who discovers the trick of cracking a nut with a stone will teach her troop. Decades later, her descendants will still crack nuts as she did; no faster, no finer. A human, upon solving the same puzzle, scribbles the method on parchment and passes it to a child. That child's grandchild, generations down the line, stands beside a roaring hydraulic nutcracker, watching ten tons of shells split in an hour.

Same starting point. Two hands, one stubborn nut. What changes is the tempo of inheritance. For humans, knowledge accelerates across the years. It snowballs from campfire whispers to industrial thunder.

Archaeological evidence indicates that cumulative culture originated in the Middle Pleistocene, roughly 600,000 years ago. An analysis of stone artifacts spanning 3.3 million years of ancient hominin toolmaking counted the procedural steps required to produce each tool. They looked specifically at how many steps it took to make each tool, almost like following a recipe. Reading the archaeological record this way is like leafing through a sequence of cookbooks written by progressively more ambitious kitchens.

Tools from 3.3 to 1.8 million years ago required only 2 to 4 moves. The recipe was brief. Smash one stone against another. Collect the chipped shard. By the period ending around 600,000 years ago, tools required four to seven steps. The same level of complexity a non-human ape might achieve, or a modern human without guidance. The recipe is lengthier now, yet remains accessible to the untrained hand.

Then, the line leaped forward. Tools now demanded more than ten intricate steps. Each builds on the last. The cookbook swelled with specialized techniques—secrets and tricks no single mind could dream up alone in a lifetime. By 300,000 years ago, the technology crafted by human ancestors had become twice as complex as anything a chimpanzee could imagine.

The cookbook is thickening into volumes. Each chapter references earlier ones, and every recipe assumes the reader's growing mastery. This leap traced the birth of cumulative culture. Knowledge is preserved, elaborated, and handed down. Each generation adds new pages to the story.

Language and complex toolmaking lean on similar mental moves. The way we build sentences inside sentences, one idea nested within another, mirrors the way we build complex tools by nesting one procedure inside another. The two abilities may have grown up together. Each scaffolding the other.

From that prehistoric inflection point on, cumulative culture became our chief advantage. Writing formalized and extended it. The printing press distributed it. The internet made it globally accessible. Each shift changed what could be known and how it could be searched, combined, and applied. None of these steps created knowledge on its own. Each fundamentally changed how human knowledge could flow.

Put simply: each earlier transition built a bigger warehouse, adding new shelves and endless aisles. Still, the job of the human curator remained unchanged. It was people who wandered the stacks, giving names to the objects, deciding their meaning, arranging them in patterns that made sense—order imposed on abundance.

## WHEN MACHINE SYSTEMS ENTER THE INFRASTRUCTURE

Large language models are a different kind of transition. Unlike previous systems that stored what humans had written or helped humans search through it, these models generate new text on demand by drawing on patterns learned across the entire record of human language.

To grasp the scale, picture every book the Library of Congress has ever shelved, every academic paper ever composed, every blog post ever tapped into existence, every Wikipedia entry in every language, every recipe, every legal ruling, every patent ever filed. Now imagine all of it devoured by a single mind. This mind never sleeps, never forgets, and responds to any question before you can draw your next breath.

It has read more in a year than every human alive has read in a lifetime, combined. It does not understand any of it the way a human reader understands a book. It retains the statistical shadow the writing casts. The shadow is enough to produce content that looks like it came from someone who did understand.

When this generative capacity is folded into operating systems, search engines, email clients, and productivity software, it ceases to be a discrete tool and becomes part of the cognitive environment. The boundary between what a human and what a machine produce becomes harder to see.

In 2024, Google began serving machine-generated responses to more than a billion search users. Some of the results were embarrassing. Suddenly, some of the results were more than just wrong; they were bizarre or even dangerous. The system even picked up internet forum jokes, treating them as real advice. It told people to put glue on pizza to keep the cheese from sliding off. Medical misinformation was also a regular occurrence. It even recommended that people with kidney stones drink their own urine. It did this with no real scientific basis—just internet rumors.

The problem was that there weren't clear signs to indicate whether an answer came from a machine or a real person. So, if you searched for something important, you couldn't easily tell if the advice was trustworthy or made up by an algorithm that didn't really know what it was talking about.

The more consequential hidden layer is older and less discussed. Long before generative systems appeared on iPhones and search bars, algorithms quietly threaded themselves through daily life, camouflaged in routine.

Picture an ordinary Tuesday: your phone greets you at breakfast with headlines chosen just for you. An ad flickers into view, tailored to your recent searches. A credit card transaction sails through or is stopped cold the instant you land in a foreign airport. A social post slips away from your feed before you can finish reading the headline. Invisible hands sort, filter, and decide. Each optimizes for a purpose you never chose. They shape your day while you barely notice.

**Recommender systems** decide which news articles, videos, and social posts appear to each user. They are tuned by engagement data collected over years of interaction. They learn that outrage holds attention longer than calm.

**Fraud detection systems** approve or block financial transactions in milliseconds by analyzing patterns across millions of accounts. The same system that quietly waves a card through at a grocery store will decline it the first time the cardholder tries to buy coffee in Lisbon, because Lisbon is not in the pattern.

**Algorithmic trading systems** execute securities orders at speeds no human can match. They react to market signals that no human reads in real time. By the time a human trader has finished reading a Federal Reserve announcement, several billion dollars of stock have already changed hands based on how the algorithm parsed the wording.

**Content moderation systems** decide which posts stay visible and which disappear across platforms with billions of users. They are the largest editorial force in human history. The editors are not human, and most of them have never been seen by the writers whose work they accept or reject.

None of these systems step into the spotlight. They are the hidden scaffolding of daily life. They are as unseen as the water coursing under city streets, as ever-present as the current humming behind drywall. Most people notice the water main only when a geyser erupts in the road, the electricity only when the room goes dark. These systems are felt most acutely in their absence, their presence a background hum we register only when it falters.

What sets these systems apart from older forms of automation is opacity. A thermostat follows a rule a person set. "Keep the room at sixty-eight." A calculator performs an operation a person started. "Add these two numbers." An algorithmic recommender optimizes an objective the user never chose, toward outcomes the user often cannot identify.

The system is not following instructions. It is pursuing a learned objective shaped by the data, the organization's incentives, and feedback signals from aggregate user behavior. Individual users see the output. They have no view of the objective, the training data, or the incentive structure shaping what they see.

The problem is not that the technology is malicious. The problem is that the technology is indifferent to the distinction between accuracy and plausibility. A language model optimizes for the next token that fits the pattern. Factual accuracy is not a training objective. Where the training data is accurate and well-represented, outputs tend to be accurate. Where the training data is biased, sparse, or contested, outputs reflect those defects without flagging them.

The system is like a fluent traveler who has read every guidebook to a country but never set foot in it. The descriptions are confident, detailed, and sometimes invented. The traveler cannot tell the difference because they have no way to check.

## THE THREE ILLUSIONS

As machine systems become increasingly embedded in our infrastructure, a cascade of cognitive risks emerges. These risks share a familiar silhouette. Each one swaps the true sensation of understanding for its polished imitation, trading substance for the illusion of knowledge.

**The illusion of explanatory depth**. When a person in a community claims to know something, others tend to assume they share that knowledge. In machine systems, fluent and confident answers are taken as evidence of the user's authentic understanding, and sometimes by observers as well.

We've lived with this illusion long before machines. Most people will confidently claim they know how a flush toilet works—until they're handed a pencil and asked to sketch the mechanism. The tank, the siphon, the chain are all suddenly elusive, despite years of proximity.

Chatbot summaries now scale this illusion to the industrial level. Ask someone who's used a chatbot to summarize a dense academic paper if they understand it, and the answer is almost always yes. Ask them to reconstruct a single argument from memory, without the machine's help, and the certainty dissolves. What was once vague familiarity now stands revealed as a mirage.

**The illusion of exploratory breadth.** The questions that

machine systems handle well start to feel like the questions worth asking. Questions the systems cannot address slip down the agenda. This pattern has a name. It is called the streetlight effect, a reference to an old joke.

A man loses his keys at night and searches for them under a streetlight because that's where he can see. The spot where the keys actually fell is hidden in the dark. Yet, the man keeps looking in the pool of light because it's easier. Sometimes, we do the same thing with questions and answers. We focus only on what's easy to find, even when the real solution is out of sight.

**The illusion of objectivity.** Machine systems appear to take an impartial stance, either because they represent every perspective equally or because they hold none. The systems are built by humans who make specific choices about training data, architecture, and objectives. Those choices encode a standpoint. The standpoint is hidden in plain sight, baked into the patterns the model learned and the metric it was trained to maximize.

An example of this is a world map. Greenland often seems almost as big as Africa, even though Africa is actually about 14 times larger. This is because the way we choose to draw—or "project"—the map changes how we see the world. Every map projection has a point of view, and that perspective shapes what we see. There's no such thing as a completely neutral map, because every choice in making the map influences how the world looks to us.

The idea that how we represent things shapes what we notice and what we miss doesn't apply only to maps. It pops up in many places, sometimes with bigger consequences. One of the most important examples is the way we depend on single solutions, or "monocultures." History shows that putting all our trust in one approach can make everything seem stable... until suddenly, it isn't.

## THE MONOCULTURE PROBLEM

Monoculture is a word borrowed from agriculture. The canonical case against monoculture is that of Ireland in the 1840s. Nearly every meal depended on one humble crop: the Lumper potato. Fields across the countryside were thick with these potatoes. They were easy to grow and filling. For many families, they were the difference between hunger and survival.

Then, an invisible enemy arrived on ships from the Americas: a plant disease called *Phytophthora infestans*. The blight swept through the fields. It turned healthy green plants into black, rotting stems almost overnight. The Lumper had no defense against this invader. Within months, the food supply collapsed.

Around a million people died of starvation. Millions more were forced to leave their homes. It became a harsh lesson written into history. Relying on just one solution, no matter how efficient, can be catastrophic when everything depends on it.

Scientific monocultures fail in many places at once. Imagine if every class in the world used the exact same textbook, written by just one group of teachers on one subject, and followed only one way of thinking. If there's a mistake in that book, everyone learns the same mistake. It spreads everywhere. That's what happens when one machine system, trained on one set of information and one way of doing things, is used everywhere. Any error or bias in that system can quietly spread into every part of science or learning, because there aren't enough different viewpoints to catch or fix it. Instead of mistakes staying small, they show up everywhere at the same time.

Two forms of monoculture compound each other.

**The monoculture of knowing.** Machine tools are like magnets for certain kinds of questions. They quickly solve the problems for which they're built. Researchers naturally start asking only the questions these tools can answer, as we saw with the illusion of exploratory breadth. This slow shift is like a city that gradually grows around its roads. At first, there are lots of ways to get around. Over time, every street and building is designed for cars. No one ever announces, "Let's make this a city just for vehicles!" Instead, it happens through thousands of small decisions, each one making it a little easier to drive and a little harder to walk or bike. Eventually, the old ways fade from view, and the alternative paths become nearly impossible to find.

**The monoculture of the knower.** Machine tools are beginning to substitute for human thinkers in certain roles. They are draining off the variety of perspectives that catch errors and open unexpected paths. Scientific progress has repeatedly leaned on investigators with different backgrounds and cognitive styles approaching the same problem from incompatible angles. The structure of DNA arrived in part from a chemist, a physicist, and a crystallographer arguing about the same evidence from different starting points. Centralizing the cognitive infrastructure of knowledge production removes much of that friction, and most of its benefits along with it. A lab in which every member uses the same chatbot to summarize the literature is a lab with one reader, no matter how many chairs sit around the table.

The issues with monoculture and sameness in science and technology aren't about crops or even about ideas. They're about how we design the very systems that power our world. When everything starts to look and work the same, it changes more than what we know. It changes how we solve problems. This shift sets the stage for another important change: the move from clear, step-by-step rules to systems that learn and optimize on their own.

## FROM RULES TO OPTIMIZATION

The shift from rule-based automation to learned optimization has a shorter history than most people expect. Aircraft autopilot systems introduced in the 1930s followed clear, step-by-step instructions called control laws. If a plane drifted from its set altitude, the system would make a precise adjustment to correct it.

Cruise control in cars works in a similar fashion. It automatically maintains speed by following simple, predictable rules. Anti-lock braking systems, which became common in the 1970s, depend on sensor thresholds and pre-programmed cycles to prevent skidding. All these systems are basically rule-following machines. They execute instructions crafted by human engineers and checked by regulators. The rules are written down in manuals, so if something goes wrong, an investigator could flip through the pages, find the exact instruction, and figure out what happened.

Everything changed with the rise of large-scale machine learning. Now, instead of following fixed rules, systems are trained to chase a target—optimizing a score or metric—without a single instruction pointing the way. Their decisions emerge from the complex mix of their structure, the data they learn from, and the goals set by their designers. You can inspect each piece on its own, but the way they interact produces unpredictable behaviors that no one person planned or can fully trace back to its source.

When a traditional autopilot fails, investigators can usually pinpoint the exact rule or line of code that caused the failure. They can trace the mistake to a clear cause. When a machine learning system messes up, the error is rarely tied to one part. The behavior is spread out across millions of learned connections, shaped by exposure to billions of data points. The system's inner workings are a black box by design. This isn't because the way these systems learn makes them hard to untangle.

A rule-based system is a recipe. A learned system is a chef who has stopped following the recipe and now cooks by instinct. The recipe can be checked line by line. The chef's work can only be tasted, and tasted again, and trusted to a degree that depends on how the previous meals turned out. If the recipe makes a bad dish, you find the line where the salt went wrong. If the chef makes a bad dish, you do not find what went wrong. You eat it. You stop ordering. You decide the chef had a bad night.

## THE SYNTHETIC DATA PROBLEM

A related concern emerges from the dynamics of large-scale training. As generative models produce more text, that content accumulates on the internet. The next generation of models is trained on data that includes substantial amounts of machine-generated content alongside human-generated content. How much this matters depends on how much synthetic data is used, whether it comes from models in the same family or from different ones, and whether the patterns in synthetic data accurately reflect reality or distort it.

When too much of a model's training data comes from other models just like it, its performance starts to break down. Nature has already shown us how this story plays out. Think of a small island where animals are cut off from the wider world. With no fresh genes coming in, each generation is forced to mix from the same shallow pool. Over time, odd and sometimes harmful traits appear—like the famously jutting Habsburg jaw that spread through European royalty. The genetic code keeps looping back on itself, growing narrower and less adaptable. In much the same way, a model fed its own outputs again and again becomes stuck, recycling its own blind spots and mistakes, until it can no longer evolve or improve.

Watermarking machine-generated content, so it can be excluded from or appropriately weighted in future training corpora, is technically feasible. Whether it is systematically implemented is a separate question. The answer requires technical standards and policy incentives that are not yet in place.

All of these questions about synthetic data—where it comes from, how it's marked, and whether we can keep track of it—aren't just academic or technical puzzles. They have real-life impact. They shape the systems and decisions that touch our daily routines. As machine-generated content flows into search engines, social feeds, and databases, it becomes part of the information used to make choices about people's lives. This blurring of the line between real and synthetic sets the stage for new challenges in privacy, accountability, and trust.

## CONSEQUENTIAL DECISIONS, INVISIBLE CRITERIA

Consider a regular day for an ordinary young adult. Before they even open their eyes, their phone has already sifted through the world's headlines and chosen which news will greet them at breakfast. On the drive to work, invisible sensors inside their car measure how quickly they accelerate, quietly relaying this data to their insurance company. Without warning, their premium climbs as the algorithm now tracks every tap of the gas pedal.

Later that day, our ordinary adult sits at their desk, polishing their résumé and hoping for a promotion. Before a human ever lays eyes on their application, it's scanned by an unseen algorithm that sifts through thousands of résumés in seconds, searching for patterns it learned from previous hiring decisions. The system claims to ignore protected categories like race and sex... still, the software has quietly absorbed the biases of the past. A résumé mentioning a women's college might drift to the bottom of the pile, edged out by one listing a coed school. This is not by design. It's simply the invisible gravity of old decisions, repeating themselves through code.

That evening, the same person tries to refinance their mortgage. They fill out the forms. Meanwhile, somewhere in a server farm, a credit-scoring model crunches the data and spits out a number that will decide their interest rate. No one explains how the number was made. The model looks at zip codes. These, at times, echo the patterns of segregation. The result is lending decisions that trace the old borders of discrimination, even as the system insists it's blind to them. What feels like a neutral process is often the shadow of history, replayed through algorithms that never learned the difference.

None of these systems explains itself. None offers an appeal. By the time the person has gone to bed, a dozen consequential judgments have been rendered against them by these mechanisms. In each case, the decision is routed through a system the affected person cannot inspect, contest, or appeal through any accessible mechanism.

Traditional administrative decisions, like loan denials, insurance denials, and hiring screens, were made by humans who could, in principle, explain their reasoning and be questioned about it. They could also be held to account if their reasoning violated law or policy. There was a person on the other side of the desk. The desk had a name on it. The name had a supervisor.

Algorithmic decisions are made by systems that produce outputs without explanation and whose internal logic is not available to the person affected, to regulators, or, often, to the organizations that deploy them. The desk has been replaced by an automated process. The name has been replaced by a confidence score. The supervisor has been replaced by a quarterly report on model accuracy.

The asymmetry is the heart of the problem. A person whose loan is denied wants to know why. The institution that denied the loan can describe the system's overall accuracy. Neither answer responds to the other. The conversation that traditional administrative law assumed, between a decision and a justification, has no equivalent in the new architecture.

## WHAT CUMULATIVE CULTURE NOW INCLUDES

Recall our discussion of cumulative knowledge from before. Now, imagine living hundreds of thousands of years ago, when early humans chipped stone to make simple tools. Over countless generations, people slowly learned new tricks and passed them down by showing, copying, and practicing together. Progress was slow, like building a tower brick by brick.

Then language arrived, and suddenly people could share ideas much faster—no more relying on silent demonstration. Writing sped things up even more. Knowledge now travels across time and distance, so you can learn from someone you'd never meet. Today, machine systems are pushing this acceleration even further. Yet, unlike language or writing, machines change how knowledge grows—not just how fast it spreads.

Previous accelerations expanded the amount that could be stored and transmitted. They did not change who was doing the writing. A book contained what a human had written. A search engine returned what humans had published. Generative machine systems produce new content from learned patterns, without a human creator, claiming no particular authorship.

The content enters the cultural record alongside human-generated content, indistinguishable in form, carrying no reliable marker of its origin. Imagine a library where, from one day forward, half the books on the shelves had no author and no provenance, yet were bound, titled, and shelved exactly like the others. The library still functions. The catalog still works. The question of whether what you are reading was written by anyone who knew anything has quietly become unanswerable from inside the building.

Returning to Apple's 2024 announcement. This wasn't just another app users could tap open or ignore. Instead, it introduced a new layer woven directly into the way people think and communicate with their devices—always on, always working in the background. This system read and wrote messages, organized notes, and offered suggestions in real time, guiding what users wrote and saw across every iPhone, iPad, and Mac worldwide.

The design intent was synthesis across an individual's entire digital life. Contacts, calendar, messages, and photos all read together to provide contextually relevant assistance. Whether users understand that the synthesis is being performed by a pattern-completion system rather than an understanding system is not a question the interface answers. The interface looks like a helpful colleague reading over the shoulder. The system on the other side of the interface is a statistical engine, fluent and tireless, with no opinion about the difference between true and false.

Grasping the distinction matters. A pattern-completion system trained on the largest corpus of human expression ever assembled produces outputs that are accurate, useful, and occasionally wrong in ways that are not flagged as errors. The system cannot tell these cases apart. It has no access to ground truth. It has only statistical patterns.

Recognizing this does not justify rejecting the technology. Knowing what the tool is, what it does reliably, and where it needs verification rather than trust is the prerequisite for using it well. A power saw is useful. A power saw used by someone who believes it cannot cut their fingers is dangerous in a specific and predictable way.

The generation of humans now entering secondary school will not know a world without this infrastructure. They are growing up with generative machine systems embedded in their communication tools, educational software, search interfaces, and creative environments. The cognitive habits they develop are shaped in part by what those tools do well. They are also, less visibly, shaped by what those tools fail to do. The shape of the failures may not be apparent until the habits are entrenched.

A child who learns to spell with autocorrect develops different spelling instincts than a child who does not. A child whose first encounter with a difficult question is a fluent machine answer develops a different relationship to difficulty than a child whose first encounter is with a teacher who says, *That is a hard question. Let us work it out together.*

Critical evaluation of machine-generated content, the ability to recognize fluency as distinct from accuracy, is a skill that requires deliberate cultivation. It is not absorbed by interacting with fluent systems. Interacting with fluent systems without training in their limits produces the opposite. A trained expectation that fluent output is accurate output. The same way a person who has only ever bought ripe fruit at a supermarket may not know how to pick a ripe melon in the field, a person who has only ever encountered fluent machine answers may not know how to tell a competent one from a confidently wrong one. The signs are subtle. The skill is teachable.

However, the skill is not currently being taught at anything like the scale required by the situation.

The cumulative culture that gave our species its distinctive advantage was built through skepticism as much as through trust. Each generation inherited what the previous one produced, tested it against new circumstances, and discarded or refined what failed. The process required distinguishing between received wisdom and verified knowledge.

As machine systems move into the infrastructure of knowledge transmission, keeping that distinction becomes more important and more difficult in equal measure. The technology does not make the distinction for us. Our capacity to make it remains one of the things that makes human reasoning, with all its limits, irreplaceable in managing a world that machine systems are progressively helping to describe.

The lineage we are building is not a tool, the way previous tools were. A hammer extends the arm. A printing press extends the voice. A pattern-completion system embedded in the operating system of a billion phones is a different kind of entity, integrated into the substrate of how its users think, write, and decide. We do not yet have a settled vocabulary for what that entity is.

Naming it plainly acknowledges that something has entered the world that did not exist before, that operates by its own principles, and that will keep operating whether or not we understand what we have made. The hidden layer is where the framing matters most, because it is where the framing is most easily forgotten. We do not notice the air we breathe until it changes.

✧✧✧

# CHAPTER 4
# A NEW KIND OF RESEARCH PARTNER

PICTURE YOUR DOCTOR'S OFFICE: the cold exam table paper, the awkward questions, the wait for results. Now, imagine skipping all that. Instead, somewhere in a server farm, a digital doctor is quietly at work—one who never needs to hear you describe your sore knees or listen to your heart. This doctor learns by absorbing the stories of millions. It traces the arc of every diagnosis, the echo of every prescription, the long trail of hospital visits, lab results, and everyday details that make each patient unique. From this ocean of experience, it pieces together clues to predict what might happen next in your health.

That's the idea behind a new kind of medical model, trained on health records from millions in the United Kingdom and Denmark. The model can forecast risk for more than 1,000 medical conditions. It doesn't examine patients or order tests. It simply reads the stories already written in your data, and from that, draws a map of your future risks.

This program, Delphi-2M, takes its name from the legendary oracle at Delphi. Seekers once traveled to Delphi to hear riddling glimpses of their fate. Unlike its ancient namesake, Delphi-2M offers no cryptic prophecies. Instead, it speaks in the clear language of probabilities. Powered by sophisticated algorithms, it sifts through oceans of health data, weaving together patterns to predict the chance of future medical events.

Delphi-2M mimics the work of an experienced family doctor after a forty-year career. Once a clinician has seen enough patients, certain symptom-habit combinations point to clear conclusions. The patient with poorly controlled blood pressure, a sedentary job, and a parent who died at sixty of a heart attack is a different patient from the one with the same blood pressure but no family history and a daily walk. The veteran clinician knows this from decades of cases. The model absorbs it by sifting millions of cases at once. The clinician learns one patient at a time over forty years. The model learns from millions of patients in weeks.

Health isn't simply about one disease at a time. Each condition—like high blood pressure or diabetes—is like a different ingredient in a recipe. Alone, each shapes your health in its own way. Mix them together... maybe add a history of depression, a desk job, or a past smoking habit, and suddenly the outcome can be very different, even surprising. The way these factors combine can change your health's future in ways that no single medical issue could predict.

The program maps these compound trajectories. It also generates synthetic patient records, plausible invented life histories that researchers can use without breaching any real person's privacy. It forecasts population health trends at a massive scale.

Predictions from systems like Delphi-2M may not guide individual appointments for a while yet. The technology has to be validated, regulated, woven into workflows, and accepted by clinicians and patients. Each step takes time. The more immediate use is at the population level. Public health planners, hospital networks, and insurers use such forecasts to replace blunt screening guidelines with calibrated ones.

Today's standards, colonoscopy at 50 for everyone and mammogram at 40 for everyone, are like setting every household thermostat to the same temperature. The rule ignores who lives in the house, the local climate, and the house's insulation. It overscreens those at low risk. It underscreens those at high risk. The low-risk group endures unnecessary procedures. The high-risk group needs earlier and more frequent attention than the rule allows.

A model that estimates risk individually changes this. The 40-year-old with a strong family history and a high-risk variant is screened at 35. The 70-year-old, in good health with two clean colonoscopies, stops being screened. Screens go to those most likely to benefit. Interventions step out of the way for those who would only be inconvenienced.

The arrival of tools like these is one sign that machine systems are moving from medical novelty to medical infrastructure. The movement does not happen overnight. Knowing where it stands today requires knowing where it began.

## THE LONG ROAD TO MACHINE MEDICINE

The idea of letting computers help doctors was once as futuristic as flying cars. In the late 1950s, when these first serious proposals appeared, most hospitals didn't even have a single computer humming in the basement. The notion seemed almost science fiction.

Then came 1959. Keeve Brodman and his colleagues saw diagnosis as a puzzle to be solved by logic. A doctor would gather clues—symptoms, test results—and match them to patterns of disease, a bit like a detective sorting through evidence. Brodman reasoned that if a diagnosis followed logical steps, then, in theory, a machine could follow those same steps. *Why not let computers join the case?*

Fast forward 11 years: William Schwartz imagined computers as sidekicks to doctors. In his view, these machines could boost a physician's brainpower, helping them see patterns or possibilities that might otherwise be missed. It was a vision both bold and strangely prescient—a glimpse of a future that was closer than anyone guessed. To see how those predictions played out, it helps to look at the earliest approach: rule-based models.

**Rule-based models** assume that expert knowledge consists of many independent, situation-specific rules. String the rules together in chains of deduction, and the result simulates expert reasoning. If the patient has chest pain, shortness of breath, and a history of smoking, then consider pulmonary embolism.

These systems were like students who aced the exam when every question matched the textbook, but stumbled when faced with the real-world curveballs. The most famous example, MYCIN, was created at Stanford in the 1970s. MYCIN could diagnose bacterial blood infections as well as the best faculty experts—at least on paper. Yet, it never set foot in a real clinic. The system didn't fit easily into doctors' daily routines, and no one could say who would be blamed if its advice led to a mistake. While rule-based systems like MYCIN relied on codified expert knowledge, another class of computer-aided diagnosis grew up alongside it.

**Pattern-matching systems** align a patient's symptoms with a catalog of illness blueprints. It's like a diagnosis based on matching fingerprints, not on understanding a patient's personal story. The system itself has no grasp of how diseases actually work. It can't explain that someone is dangerously low on blood because she suffered three days of bleeding. All it can do is declare, "*These symptoms look like this illness I've seen before.*"

When the match is close, the diagnosis is spot on. Yet, if the fit is imperfect or the patient's story is unusual, the system has nothing to offer. The main problem with these early computer systems was that they didn't really understand how diseases work. They could follow instructions or spot patterns. They couldn't figure out what was actually happening inside the body.

In the 1980s, computers just weren't powerful enough to make smarter systems work for real patients. Raw computing power was not the only limit. Medicine also needed new data, in new forms, at new scales.

The 1990s marked a turning point. The world's medical records began to leap from dusty file folders into the digital age. By the time the mid-2010s arrived, hospitals from Chicago to Shanghai had traded their paper mountains for sleek databases. Patient histories, once hidden away in basement archives, became searchable maps. They were ready for doctors and researchers to trace patterns, spot trends, and ask new questions that no one had asked before.

Meanwhile, the genetic revolution was quietly rewriting what was possible. When the Human Genome Project began, scientists needed 13 years and $3 billion just to read a single person's DNA. Now, a complete genome can be decoded in less than a day for about the price of a new smartphone. The speed and affordability of this change are nearly unmatched.

The data boom didn't stop there. Medical images and wearable devices began flooding the field with information. Hospitals now collect thousands of scans every year. Each one is a digital treasure chest, packed with details so dense that even expert eyes can't see it all. At the same time, fitness trackers and health apps quietly log our steps, heartbeats, and hours of sleep, recording a living, breathing portrait of our health in real time.

Social and environmental data add context to this information. Air quality, the level of pollutants in the air. Neighborhood characteristics, including housing conditions and access to resources. Social network structure, the patterns and strengths of relationships among a person's connections. Shopping patterns, the frequency and types of purchases. All of these carry health information that was previously invisible to clinicians.

However, the data is often unstructured and chaotic. Turning it into something useful required new mathematical methods. Transformer architectures, the same family of models behind large language models, assign different weights to different parts of an input rather than treating them equally. Neural networks grow deeper and more capable as computing power enables real-time output. The stage is set for medicine to use a generation of tools originally built for other purposes.

## THE RISE OF THE AGENTIC SYSTEM

Machine systems, in their earliest clinical forms, were narrow in scope. They performed a single task on a single data type and required human monitoring at every step. The machine flagged. The clinician decided. The division of labor matched the tools' capabilities. It also set the regulatory and cultural expectations that still hold in medicine today.

What has changed recently is the emergence of what is called agentic systems. These are machines that process information or generate content. They also initiate, plan, and execute sequences of actions toward open-ended goals, adapting as they learn. The earlier generation of clinical software responds within fixed boundaries. An agentic system decides on the most effective path to an outcome, sequences steps, interprets setbacks, and moves beyond static instructions. It functions less like a calculator and more like a colleague who can be given a goal and, within limits, trusted to figure out how to reach it.

The practical difference is scale and flexibility. Rule-based systems require programmers to anticipate every situation the program will encounter. Agentic systems learn from data and adjust accordingly. Their learning capacity allows them to generalize. They apply knowledge picked up in one context to a new one in a way that earlier clinical decision-support software cannot. The same capacity means their behavior in genuinely novel situations is harder to predict, which is precisely why the question of human monitoring becomes more urgent as these systems grow more capable.

Consider an office analogy. In the corner sits an old-fashioned assistant, precise and methodical. Give them a slip of paper with the file name and cabinet location, and they'll retrieve it without a hitch. Their work is predictable, their boundaries clear. Now, meet their counterpart: a new assistant with an eye for the bigger picture.

Task them with preparing a comprehensive report on the last decade's trends, and they spring into action. They search archives, explore databases and tie together findings from scattered corners of the office. They choose keywords, weigh which discoveries matter, and arrange the results into a coherent narrative. Sometimes, they exceed every expectation, surprising with insights no one thought to look for. Other times, they stumble. Their creative leaps require a careful final review. Two assistants. One follows instructions to the letter. The other interprets, plans, and adapts, bringing a touch of unpredictability—and possibility—to the familiar rhythm of the workplace.

The leap from office assistant to research partner is no longer theoretical. At the Agents4Science conference in 2024, the scale of what agentic systems can now do in research settings was on open display. Machine systems acted as lead investigator, data analyst, draft author, and peer reviewer, all in a single experiment. The systems took on every phase of the research cycle. They generated hypotheses, analyzed datasets, produced manuscripts, and assigned preliminary scores to papers alongside human subject-matter experts.

The "agents" moved through thousands of manuscripts at speeds no human or team could realistically match. Patterns rose from the data like constellations, and insights were distilled at an incredible pace. Still, when the work called for discerning genuine novelty, probing intellectual rigor, or spotting a methodological flaw hidden in plain sight, seasoned human experts remained indispensable.

The result? The most high-rated papers emerged from a hybrid approach. Machine-generated drafts, honed and critiqued by human hands. It wasn't the speed of the machine or the judgment of the person alone that yielded excellence. It was their collaboration. In medicine, too, the promise of agentic systems lies not in replacing clinicians and researchers. It lies in forging true partnerships. Each complements the other's strengths. The machine races through the data. The human brings discernment, context, and care. Yet even as these partnerships flourish, they reveal a new challenge—one rooted in the sheer scale and speed of modern science.

## SCALE, SPEED, & THE LITERATURE PROBLEM

Global research output now doubles every 9 years. Each year, more than 5 million new articles are published. Science is coming at a pace of one paper every 6 seconds, around the clock, every single day.

A diligent clinician, faithfully reading 10 papers a day, every weekday, all year, might finish 2,500 papers. That's a drop in the ocean. In just one specialty, like neurology, oncology or cardiology, the deluge is overwhelming. The literature outpaces a lifetime's worth of reading. The flood is not just daunting. It is becoming a structural problem.

Important findings go unnoticed because nobody has read the paper. Promising drug candidates are not pursued because the relevant preclinical work was published in an adjacent field nobody on the team subscribes to. Clinical trials are designed without knowledge of failed approaches already in the literature. The same trial that failed in 2014 in Sweden gets repeated in 2026 in California, with the same negative result, because the principal investigator did not encounter the earlier paper, and the search engine used did not flag it.

Agentic tools address the bottleneck directly. Systems such as Anara sweep through academic databases, preprints, and recorded conference talks, adjusting search parameters and clustering findings for synthesis. What used to require teams of specialists working for months produces comprehensive knowledge bases in hours. This allows faster innovation cycles and reduces the risk of missed insights.

Consider a team of three librarians and a clinical fellow, trying to answer a major medical question. They need to do a "systematic review," which means searching through hundreds or even thousands of research papers on a surgical technique. Doing this the traditional way? It takes them six weeks of careful searching, sorting, and checking. With agentic tools, the same team can get a draft of the review in just 4 days. The librarians and the fellow still check the results. Now their job is much quicker. They're fine-tuning and confirming rather than starting from scratch. This partnership between humans and technology saves time, reduces busywork, and lets experts focus on the important decisions.

In the lab, the same parallel-processing capacity applies. Agentic systems examine large experimental datasets, identify outliers and recommend alternative analyses. They also often highlight patterns that manual inspection misses. In genomics and epidemiology, these agents automate analytical tasks in parallel, running dozens of comparisons at once. Agentic systems also design and execute follow-up. They adjust parameters and improve reproducibility. This frees experts to focus on strategic questions that require human reasoning.

The acceleration is real, though not without limits. Current generative systems perform well on routine problems in familiar domains and struggle with novel or highly complex situations. DeepMind, OpenAI, and Meta are pursuing different approaches to building more robust models. These include internal representations that let a system reason about unfamiliar environments rather than pattern-match against what it has seen before. Whether any of these approaches will produce clinical-grade reasoning in open-ended diagnostic situations remains to be seen. The tools available today are broad in scope and fast, though they remain dependent on the quality and breadth of their training data.

As the field races forward and these tools become more capable, a new question comes into focus: *what happens when the limits of technology collide with the complexity and diversity of the real world?* The next frontier isn't just about speed or scale. It's about recognizing where even the smartest machines can falter, and why.

## THE LIMITS OF THE MACHINE IN THE CLINIC

Machine learning algorithms learn from the data used to train them. If the data carries existing inequities in who receives which diagnoses, which populations are represented in clinical datasets, and which diseases are overrepresented in the literature, the algorithm learns those inequities as if they were facts about the world. A model trained predominantly on data from affluent urban hospitals performs poorly in rural or low-income settings. A model built on data that underrepresents women or people of color makes less dependable predictions for those groups.

To see how real-world bias can sneak into healthcare technology, consider a major 2019 study published in *Science*. It looked at a popular algorithm used by a large U.S. health system. This algorithm was designed to help doctors find patients who needed extra support for serious health problems.

Instead of looking directly at how sick a patient was, the algorithm used how much money had been spent on their healthcare in the past as a shortcut to determine "how sick they were." Here's the problem: in the United States, Black patients have often faced barriers to getting care, so they typically spend less on healthcare—even if they were just as sick as White patients. The algorithm didn't "see" these barriers. It only saw the numbers.

Due to this flaw, Black patients had to be much sicker than White patients before the algorithm would recommend them for extra care. The program never asked about a patient's race. Still, by using spending as a stand-in for sickness, it accidentally built in bias from the real world.

This pattern, in which algorithms reflect and perpetuate biases present in their training data, does not remain confined to care management. It extends to other domains as well. Machine learning is now an accepted practice for reading electrocardiograms, plain radiographs, CT and MRI scans, skin images, and retinal photographs. In these applications, algorithms assist clinicians by flagging variations from the norm.

*Whose norm is embedded in the model?* If the training set overrepresents one demographic group, the model's baseline becomes that group's typical presentation, and important differences in others may be missed.

The clearest illustration is dermatology. Most early skin-image datasets used to train cancer-detection algorithms were assembled from clinics in Australia, Western Europe, and North America. The patients in the photographs were overwhelmingly light-skinned. The resulting algorithms are highly accurate at detecting melanoma on light skin and substantially less accurate on dark skin. The lesion in a dark-skinned patient that a dermatologist flags in person is missed by an algorithm that has never seen enough dark-skinned examples to learn what a melanoma looks like in that population. Better, more representative training data must be assembled. The work costs money and effort that institutions are not always willing to spend.

The challenge of ensuring that new tools are fair and work for everyone, not just those most represented in the original data, reappears again and again in medicine. Even as smarter systems emerge, they can still inherit the same blind spots as earlier models unless we take steps to address them.

Google's Health Acoustic Representations system, called HeAR, analyzes cough sounds to detect tuberculosis. The concept gained momentum during the COVID-19 pandemic, when audio recordings of coughs, labeled with the health status of the person who produced them, were used to build models that could detect signs of respiratory disease without active patient participation. The system is built on a massive dataset of labeled audio clips, making it more generalizable than its predecessors. The tool can be fine-tuned for multiple tasks and made available to independent researchers for validation.

Even the most advanced tools still face the messy reality of the world outside the lab. Behind every data point is a person in a real place. Environments may be quiet or chaotic, familiar or far-flung. That's where the real test begins.

One patient sits in a calm, carpeted doctor's office in San Francisco, their cough echoing in the quiet. Another coughs into a recorder at a bustling, open-air clinic in rural Kenya, where background noise swirls—children laughing, motorbikes buzzing, wind rushing past. Both may have tuberculosis, but the software's confidence depends on whether it has "heard" coughs like theirs before. The more varied and real its training, the better it can recognize illness in the wild.

The life2vec model illustrates the same power and the same problem. Life2vec is a program designed to look for patterns in people's lives and predict important outcomes. It takes all the events in a person's life, like jobs they've held, medical diagnoses, how much they've earned, or times they've been in the hospital, and turns them into a story. Each event is a sentence, and each sequence of events is a paragraph.

The program was trained on detailed life histories of millions of Danish citizens spanning the years 2008 to 2016. After "reading" all these life stories, the model learned to predict whether a person would die before 2020 with about 80% accuracy. It also found patterns. For example, people with lower incomes, mental health diagnoses, or who were male were more likely to die earlier. Life2vec could even estimate personality traits that are linked to later-life outcomes.

The program's accuracy is striking. However, the patterns may not hold for non-Danish populations. Denmark has universal healthcare, comparatively low income inequality, and consistently maintained registers. A model trained on Danish lives is, in a real sense, a model of Danish lives. Applied to Mississippi, it may not work the same way. Applied to Lagos, it may not work at all. That the model performs well in its training context does not indicate it will perform well elsewhere, and the assumption that it will is a common error in the early launch of these tools.

This challenge highlights a crucial lesson. Context matters. Health, illness, and medicine look different from place to place. That's why every new breakthrough has to be tested, adapted, and sometimes completely rethought as it moves from one environment to the next.

## READING THE BODY WITHOUT TOUCHING IT

Until recently, many medical tests required patients to actively participate or relied on specialized equipment that was not available in many clinical settings. Testing often requires trained staff, dedicated devices, and patient effort. Now, new tools, particularly deep learning models trained on large clinical datasets, can extract clinically meaningful information from widely available tests that were not originally designed for those purposes.

Advanced programs can now estimate how well a person's lungs work from a standard chest X-ray. Traditionally, that information comes from spirometry. This is a breathing test that measures airflow and lung capacity. It can be difficult for young children, older adults, or patients who are acutely ill, and it is not routinely available in many low-resource settings.

In recent studies, deep learning models trained on paired X-ray and spirometry data have been able to predict measures such as forced expiratory volume. It also detected patterns consistent with chronic obstructive pulmonary disease or COPD. Chest X-rays are already among the most common imaging tests worldwide. These models allow clinicians to estimate lung function opportunistically—using images taken for entirely different purposes, with no additional time, cost, or patient burden.

Programs can now spot other health problems on chest X-rays, too. For example, they can find signs of heart disease, diabetes, and how well the heart is pumping. Some models, like those developed at Stanford, can analyze a chest X-ray to estimate the strength of the heart. Other studies have shown that small details in an X-ray, like the pattern of blood vessels or the thickness of tissues, can help predict if someone is at risk for heart failure or even dying early. These predictions help doctors decide who might need more tests or extra care, all from a single, simple X-ray.

Other types of images are becoming more useful as well. A photograph of the retina has long been used to diagnose conditions like diabetic retinopathy. Now, these photos help spot health problems throughout the whole body. For example, teams at Google Health and Moorfields Eye Hospital discovered that images of the retina can help predict a person's risk for heart disease, high blood pressure, smoking habits, and even major heart attacks. Other studies have found that certain features in eye photos can be clues for kidney disease or the early stages of brain conditions like Alzheimer's.

Together, these approaches reflect a broader shift in medicine. Instead of ordering new tests, clinicians can increasingly "read more" from the tests they already have. This expands access to screening and early detection, particularly in settings where specialized diagnostics are limited.

These advances are still just one part of a much bigger story. The same powerful tools are now being used in entirely different fields. One of the most exciting frontiers is genomics—the study of our genes and how they work. Here, computers are helping unlock the language of DNA, revealing new insights into health and disease that were once hidden deep inside our cells.

## THE GENOMIC LANGUAGE

DNA is a language with its own grammar and rules. Genes are like sentences, where the order and context of each word shape the meaning. The four-letter alphabet of DNA—A, C, G, and T—writes these sentences, layering information on top of information. Every detail matters. Which proteins are made, when they're made, how long they last, when the genome copies itself, and how the cell's library of instructions stays organized are all written in sequences only a cell can read.

Surprisingly, only about 2% of our DNA actually codes for proteins. These are the main building blocks of our bodies. Think of constructing a living system as building with Lego. The proteins are the bricks. The instruction booklet tells you which bricks to use, in what order, and what to build.

A liver cell and a brain cell both have the same box of Lego pieces, the same DNA. Yet, they follow completely different instructions. One filters blood. The other carries thoughts. Even two very different creatures can share most of the same Lego bricks, yet look and act nothing alike. This is due to their instruction booklets. Their genetic "stories" are different. The secret patterns that write these stories are subtle, ever-changing, and only partly understood.

To tackle these complex genetic puzzles, programs called "foundation models" are used. These models are first trained on huge collections of DNA sequences—without anyone having to label or explain the data. After this first step, the models are adjusted to solve specific problems, such as identifying which genes are turned on or off, where certain proteins bind to DNA, or what parts of the genome do what jobs. This approach has helped machines do much better at understanding and predicting how our genes work.

GROVER, an acronym for Genome Rules Obtained Via Extracted Representations, is one such foundation DNA language model. The program is trained with a vocabulary derived from the human genome. It uses a technique that produces a "frequency-balanced vocabulary" by repeatedly merging the most common adjacent pairs of letters into single tokens. The same technique underlies the tokens used in large language models for human languages, as we saw in the earlier chapters.

Applied to DNA, the program produces a vocabulary that captures what the genome treats as a meaningful unit, rather than imposing a preselected vocabulary. GROVER learns sequence context, structural rules, and lexical ambiguity in genomic sequences. On tasks involving genome element identification and protein-DNA binding, it outperforms earlier models.

Treating the genome as a language to be learned rather than a code to be manually deciphered makes it accessible to analytical tools developed for natural language. The same machinery that learned the rules of English by reading the internet now learns the rules of human DNA by reading three billion base pairs. The rules are different. The mathematical model for learning them is the same. Artificial intelligence is reading the source code of biology.

All of this matters in real life because it explains how changes in a person's DNA might affect their health. Some changes, called variants, might do nothing at all. Others can cause disease by switching genes on or off, or by disrupting how proteins work. Figuring out which variants are important is a critical task. The databases are massive but still lack key pieces. That's where these new models come in: by learning from all the DNA data we have, they can make educated guesses about the effects of new or rare variants—even if those have never been studied before. This assists in mapping the human genome faster and identifies which changes might matter for real patients.

## PATHOLOGY REBORN

Among the clinical specialties most concretely altered by machine systems, pathology occupies a central position. Pathologists interpret tissue samples to identify cancer, characterize tumors, and guide treatment decisions. For most of the discipline's history, the work involved a pathologist hunched over a microscope, peering at a thin slice of tissue on a glass slide.

Digital pathology replaces the glass slide with a high-resolution digital image that can be shared, stored, and analyzed computationally. The same tissue sample is transformed into a crystal-clear, massive digital image rich in detail. Every cell nucleus visible, every structure preserved. These digital slides, sometimes a gigabyte or more, can be instantly shared around the world, stored for years, and analyzed by powerful computer programs. This leap to digital pathology brings the field into the age of machine intelligence. It enables doctors and machines to work together to spot patterns and make discoveries once impossible.

The change is comparable, in its way, to the transition from film photography to digital. Film captures an image. Digital captures an image and makes the image computable. A film photograph could be examined by an expert. A digital photograph can be examined by an expert and by a million expert eyes at once. It can be indexed, searched, compared against large libraries of similar images, and interrogated by tools that look for patterns no individual examiner has time to find.

The practical uses are ambitious. One example is VirtualMultiplexer. It is a generative tool that synthesizes images showing the distribution of multiple protein markers in a tissue sample from a single input slide image. Producing these multiplexed images experimentally requires staining a tissue sample with multiple antibodies, a process that consumes time, material, and tissue, and can degrade the sample. VirtualMultiplexer generates computationally what would otherwise require repeated laboratory processing.

Another program, PathChat, pairs a vision encoder trained on more than 100 million histology images with a large language model pre-trained on 13 billion parameters. It is also fine-tuned on more than 450,000 instructions covering diagnostic and open-ended pathology questions. PathChat outperforms previous models across all tasks tested. Adding clinical context, patient age, sex, clinical history, and radiology findings improves accuracy on multiple-choice diagnostic questions from 78% to 89%.

The improvement reveals a basic clinical truth. A pathologist who knows the patient is a 65-year-old smoker with a lung mass interprets a slide differently from one who is handed an unmarked specimen. Context guides interpretation. The model that incorporates clinical context performs better, just as the experienced pathologist with the chart in hand performs better than the trainee with only the slide.

Prov-GigaPath is another whole-slide pathology foundation model. It is pre-trained on 1.3 billion image tiles from 171,189 whole slides collected by Providence, a health network of 28 cancer centers across the United States. The approach treats individual image tiles the way a language model treats individual tokens. It learns representations of pathological features from their context within slides and their variation across the dataset. The result is a model that can be fine-tuned for a range of pathology tasks with fewer additional labeled examples than approaches that start from scratch.

The cumulative effect is that a single pathologist, with access to machine-assisted systems, analyzes tissue samples at a scale and speed previously impossible. Instead of reviewing individual slides under a microscope one at a time, the pathologist supervises a system that processes them in parallel. It flags anomalies, identifies patterns, and generates predictions about tumor characteristics that guide treatment selection. Precision oncology, which matches specific therapies to tumor molecular profiles, depends on this kind of detailed characterization. Machine tools change the process, making it faster and more consistent.

The labor implications are not trivial. The American Society of Clinical Pathologists has warned for over a decade about a shortage of trained pathologists. Machine tools that allow each pathologist to do more work do not solve the shortage. They reduce its bite. They shift the work from microscope time toward review and decision time. They change the training pathologists undergo, the daily workflow, and the structure of the specialty's contribution to patient care.

## TEACHING THE NEXT GENERATION

The arrival of these tools in clinical practice comes at a moment when medical education is actively working out how to respond.

Ranga Krishna, who developed the medical education program at Duke-NUS Medical School in Singapore and directed clinical and education programs at Rush Medical Center in Chicago, talks directly about the challenge. The model of medical education that Krishna describes builds on Kolb's taxonomy of learning styles. Each style is paired with a way machine tools support it.

**Concrete experiential learners** use machine-driven simulations and virtual environments to practice clinical skills in real-world scenarios. A medical student preparing for her first central line placement runs a simulation a dozen times before placing one in a patient.

**Reflective observational learners** benefit from evaluating large datasets, which surface patterns that support critical thinking. A resident reviewing his own diagnostic decisions across a year of cases sees, with the help of the tool, where he overdiagnoses and where he underdiagnoses.

**Abstract conceptual learners** use visualization tools to understand intricate physiological systems. A complex feedback loop in renal physiology, the kind that earlier generations of students memorized from a textbook diagram, now appears as an interactive animation that responds to the student's inputs.

**Active experimental learners** test hypotheses in simulated clinical scenarios. The student who learns by doing is given a steady supply of cases to work through, each with adjustable parameters, immediate feedback, and the option to try again.

The most experienced clinicians often combine all four styles. The case that they cannot resolve using rules or pattern-matching becomes a question they probe through simulation. The simulation leads them to a hypothesis. The hypothesis points to a test. The test sends them back to the literature. Machine tools support every stage of that loop. Students who grow up using them as learning partners may develop habits of mind their teachers did not.

The risk is that they may also develop habits their teachers would not endorse. A student who never struggles with a difficult differential diagnosis without a chatbot at her shoulder may not develop the same diagnostic instinct. A resident who outsources literature review to an agent may never learn to read a paper critically. The educational challenge is not to keep the tools out of the curriculum. It is to design a curriculum that produces clinicians who use the tools well, recognize their limits, and retain the disciplined judgment that the limits demand.

Stepping back from the details, a bigger theme emerges—one that connects every tool and every example in this chapter. That theme is partnership.

## THE PARTNERSHIP FRAME

Delphi-2M predicts. The public health planner decides what to do with the prediction. GROVER spots patterns in DNA. The geneticist asks what those patterns mean for the patient in the next room. PathChat classifies a slide. The pathologist confirms the classification and connects it to a treatment plan. HeAR flags a likely tuberculosis case. The clinic confirms with a sputum test and starts the patient on treatment. The agentic literature tool drafts a systematic review. The team verifies and edits the draft into a publication.

In every case, the machine extends what the human can do without replacing the judgment that the human brings. The partnership is asymmetric. The machine processes more data faster and never tires. The human supplies context, accountability, and the moral weight that decisions about another person's body require.

The frame matters because it sets the terms for everything that follows. A medical system that treats machine tools as oracles produces one kind of medicine. A system that treats them as colleagues who are useful but fallible produces another. The first kind drifts toward errors that the system cannot catch because it has surrendered the catching. The second kind builds in the verification that the tools themselves cannot perform.

Artificial intelligence is reading the records that human medicine has produced. It is sorting through more cases than any clinician will see in a career. It is producing predictions that are sometimes startling and sometimes wrong. The work of medicine, in the years ahead, will be to receive what it produces with the same combination of curiosity and skepticism that has always defined good clinical practice. The instruments are new. The discipline of using them well is old.

✧✧✧

## CHAPTER 5
# DRUG DISCOVERY'S DIGITAL COLLEAGUE

EVERY MEDICINE that finds its way into a patient's hand begins as a glimmer in the mind of a chemist or biologist who glimpses a problem in human biology. They imagine a molecule, slight and unseen, that might repair a broken process in a cell. That fragile hope faces a gauntlet. From thousands of possible drugs, only a rare few endure the long journey, finally appearing—years later—on pharmacy shelves.

On average, the journey to a prescription bottle is a marathon of 10 to 15 years. The cost soars between one and three billion dollars. For scale, designing and building a new commercial aircraft from scratch takes about a decade and tens of billions of dollars. Drug development's burden is shouldered by thousands of researchers across dozens of institutions.

Only about 10% of potential new medicines make it through testing to become approved drugs. Most candidates drop out at some point along the way. Sometimes, a drug that seems to work well in animals doesn't work in people. Even drugs that look safe and helpful in small human studies can show problems when tested with more people.

Drugs considered safe may turn out to be less effective than existing therapies. Drugs that do work may show troubling variability across patient populations, undermining the case for approval. Each failure costs years of effort. Each costs tens to hundreds of millions of dollars spent on a question that ends with a negative answer. The cost of every failure is built into the price of the eventual successes. That is why drugs that work are priced the way they are.

Taking it a step further, searching for a new medicine is like wading through a universe of possibilities. Estimates suggest there are at least 10^33 possible drug-like molecules. That's a 1 followed by 33 zeroes, a number so huge it's almost impossible to picture. If a group of chemists worked day and night, making a thousand new molecules every year for 40 years, they'd only create a million.

Information, not money, is the deepest constraint on drug discovery. Biological systems are complex. Billions of molecular structures could plausibly interact with any given target protein, even within reasonable size and shape limits. Traditionally, drugs are discovered by synthesizing candidate molecules, testing them, interpreting results, and iterating. The process moves step by step. It is limited by how fast people work and how quickly labs can run experiments.

Today, machine systems are like powerful new explorers in the world of drug discovery. They can handle the entire digital side of the search, from picking out which biological targets might matter to polishing the best drug candidates. The number of possible molecules is still unimaginably huge. What's changed is how we look for them.

The shift continues the story this book has been telling. The instrument that turned the lab assistant into a lead investigator, that read tissue slides at scale, that found risk patterns across millions of patient records, has now arrived at the lab bench. The collaborator who was reading papers and synthesizing literature is now proposing molecules.

## THE PROTEIN FOLDING BREAKTHROUGH

For more than 50 years, structural biology has wrestled with protein folding. The question sounds simple, yet solving it is like decoding a secret language. Imagine each protein as a long string of beads arranged in a unique sequence by the instructions in our genes. In the blink of an eye, this string twists, loops, and snaps into a complex three-dimensional sculpture.

That final shape is everything. It decides which partners the protein will dance with, which chemical reactions it will spark, where it will travel inside the cell, and how it will keep the machinery of life running smoothly. If the protein folds the wrong way, the results can be disastrous. A single misstep, and the protein might lose its job or, worse, cause harm. Diseases like Alzheimer's, Parkinson's, type 2 diabetes, and the infamous prion disorders all trace back to proteins that have lost their way in the folding process.

Understanding the shape of a disease-related protein is the essential first step to designing a drug that can change what it does. A protein's surface is a landscape filled with peaks and valleys. One of these is the perfect spot for a drug to do its work. Without a detailed map of these, making a drug to fit is like trying to craft a key for a lock you've never seen. The techniques used to uncover these protein structures are slow and delicate.

The oldest and most famous tool is X-ray crystallography. It has been used since the 1950s. First, a tiny, sparkling crystal forms from the protein. Then, powerful X-rays shoot through this crystal. When the X-rays hit, they scatter and create a unique pattern. By studying this pattern, it is possible to figure out where every atom in the protein sits. However, not every protein can be crystallized, and even when it works, understanding the pattern can take years.

Another method is cryo-electron microscopy or cryo-EM. Instead of growing crystals, this technique freezes proteins in a thin ice layer. It essentially captures them in a frozen snapshot. They are then zapped with electron beams, and super-detailed images are able to be collected. Cryo-EM can visualize proteins that refuse to form crystals. However, the machines are expensive and require highly trained experts to operate.

A third tool is called nuclear magnetic resonance or NMR spectroscopy. In this technique, powerful magnets and radio waves are used to detect signals from the nuclei of atoms. NMR is ideal for looking at small proteins. For really large proteins, this method struggles to keep up. Each of these techniques helps us unlock the secrets of proteins. They all require great skill, patience, and sometimes a lot of time.

Each determination by any of these methods often takes years of lab work. By the early 2000s, about 100,000 protein structures had been determined worldwide. A collective achievement. Also, only a small fraction of the protein universe. The human proteome alone contains about 20,000 proteins. Those proteins fold differently in different contexts. They interact with other proteins and adopt multiple shapes. The total number of relevant structures runs into the hundreds of thousands. The experimental pipeline and these tools simply cannot keep up.

The race for an alternative has been running for decades. Computational protein structure prediction, the attempt to derive structure from amino acid sequence using physical principles or learned patterns, has improved slowly and steadily. The Critical Assessment of Protein Structure Prediction competition, called CASP, has been held every two years since 1994. It scores the best predictions against newly determined experimental structures. Through the 2000s and 2010s, the best methods inched forward. None reached experimental accuracy. Solving the folding problem began to feel like one of those scientific challenges that yield slowly across generations. Better each decade. Never quite done.

Then, in late 2020, AlphaFold made a breakthrough that changed everything. Created by DeepMind, AlphaFold learned and improved through countless rounds of training. It predicts the three-dimensional shapes of proteins with astonishing precision—so close to experimental results that the gap is about the width of a single atom.

The program was trained on the Protein Data Bank, the public archive of experimentally determined protein and nucleic acid structures. It learned mapping between sequences and structures with enough precision to generate reliable predictions for proteins whose structures had never been determined experimentally. The practical consequence followed immediately. Within two years, AlphaFold generated protein structure predictions for virtually every known protein. More than 200 million of them. The database was made freely available. The full protein universe became computationally accessible for the first time.

The ripple effect on drug discovery is substantial. Where a chemist once waited—sometimes for months or years—for a crystallographer to painstakingly solve the structure of a target protein, now that structure materializes almost instantly, complete with confidence scores. Research programs that used to stall in uncertainty now surge ahead, the bottleneck of waiting for structural biology replaced by a new frontier to explore.

AlphaFold 3, released in 2024, extends the capability further. Earlier versions accurately predicted the static structure of individual proteins. AlphaFold 3 predicts folding sequences, interactions between proteins and small molecules, and interactions between proteins and nucleic acids. It also predicts the behavior of synthetic enzymes designed in the lab. The static snapshot became a dynamic picture.

For drug discovery, the difference is critical. Drugs do not bind to frozen proteins. They bind to proteins that move, flex, and interact with countless molecules in the crowded environment of a living cell. Predicting those interactions demands more than knowing where atoms sit at rest. It requires understanding *how* proteins move.

Inside living cells, proteins act differently than in isolation. Capturing those dynamic behaviors is far harder than observing the structure of purified proteins in a test tube. Some proteins, known as intrinsically disordered proteins, lack a single stable structure altogether. They flop and reorganize. They also adopt different shapes for different functions. Structure-prediction tools that excel with rigid proteins often struggle with these flexible forms, which are heavily implicated in many cancers and neurological diseases.

While scientific progress continues, many obstacles remain. The work is far from finished. With this backdrop of complexity, the next step in drug discovery comes into focus. Once a target protein is identified and mapped out, the challenge shifts. Now it's necessary to find a molecule that can interact with the protein and adjust its activity in a way that benefits patients.

## FROM TARGET TO CANDIDATE

Traditional high-throughput screening tests tens or even hundreds of thousands of compounds from chemical libraries to see which might latch onto a target and show a spark of promise. These libraries are immense, yet still limited. They are built over decades. Each compound is tied to an old experiment or a past pharmaceutical possibility.

Synthesizing and testing every molecule is painstaking work. It demands time, money, and patience. Over the years, a company's compound library becomes a living archive, shaped as much by the bold pursuits of past scientists as by the projects that never made it beyond the drawing board.

Generative models take a radically different path. Instead of fishing through the same library of old compounds, they study the language of chemistry. They focus on the rules and patterns that make a molecule drug-like. With this chemical fluency, these models invent brand-new molecular blueprints, custom-built to meet a wish list of qualities: the power to stick tightly to a target, the resilience to stay intact in the body, the gentleness to avoid causing harm.

The ideal molecule must also journey safely. It either needs to be absorbed in the gut, carried through the bloodstream, or broken down at the right pace and then sent quietly on its way without damaging vital organs. Driven by experience with millions of known drugs, these models can generate promising new candidates in a matter of minutes. Each is crafted to clear a gauntlet of scientific demands at once.

This transformation in drug discovery is similar to what happened in architecture when computers replaced pencils and drafting tables. Architects used to hunch over blueprints, painstakingly sketching each line by hand. Every new idea meant hours of erasing, redrawing, and measuring. Only a handful of building designs could ever be considered before construction began. Then came computer-aided design. Suddenly, the architect could spin out dozens of design options in the time it used to take to finish just one. The software didn't create the buildings. The architect did. The technology turned the process into a creative playground.

Drug design with generative models is now entering its own creative renaissance. Instead of laboriously inventing one molecule at a time, chemists can now sift through a vibrant gallery of proposals generated by machine partners. These are often ideas that the old, slower process could never have produced. The human still decides which molecules to pursue. The difference is that now, the field of possibilities is bigger, bolder, and more exciting than ever before.

DrugReflector and Gemma are examples of this generation of tools. The systems sift through varied patient and molecular data to identify ideal therapeutic candidates and predict how new compounds affect human cell biology. The speed advantage over classic approaches is substantial. What previously required years of iterative lab work to narrow a candidate library can now be compressed into weeks of computational screening. Lab confirmation is now simply focused on the candidates most likely to succeed.

Leading pharmaceutical companies report significant reductions in research and development costs attributable to early-stage machine-assisted research. The financial model of drug development, in which the cost of a few successes must offset the cost of many failures, changes when the failure rate in early screening drops. Fewer failed compounds advancing through costly lab stages means more resources available for the candidates worth advancing.

The improvement is real, not unlimited. The bottleneck has simply shifted. Once early-stage candidate generation is no longer the slowest part of the process, whatever step was previously second-slowest, often animal toxicology or Phase I human safety trials, becomes the new constraint. These stages require real, physical time that no amount of computational power can compress. A six-month study still takes six months.

The dream of "AI-discovered drugs" appearing in a year runs up against this stubborn reality. Biology moves at its own pace. The overall timeline from idea to medicine is growing shorter. Where 10 to 15 years was once standard, 8 to 12 years is now possible in the best-run programs. The change is meaningful, but not as dramatic as some headlines imply.

Yet there is a different, faster path that sidesteps much of the traditional process entirely. This is the field of drug repurposing. Here, machine systems in drug discovery take on a new role that doesn't require inventing new molecules. Drug repurposing, also called drug repositioning, is the search for new therapeutic uses for compounds that regulators have already approved for other indications.

## THE SHORT CUT TO REPURPOSING

Repurposing existing drugs is like discovering a hidden shortcut on a map you thought you knew. These medicines have already passed the hardest tests. Their safety is proven, their factories are up and running, and in many cases, they're off-patent, making them affordable. When a drug originally approved for one illness turns out to help with another, it offers patients a faster, less costly route to treatment than developing an entirely new molecule.

One of the most legendary stories in this field is the rise of sildenafil. In the 1980s, Pfizer was hunting for a new weapon against angina, the chest pain caused by poor blood flow to the heart. Their latest compound, sildenafil, wasn't meeting expectations. Then, something odd happens. Trial participants keep whispering about an unusual side effect and seem oddly reluctant to hand back their left-over pills. Pfizer's scientists catch the hint, change course, and give the drug a new lease on life. Thus, a failed heart medication is transformed into Viagra—a blockbuster treatment for erectile dysfunction and a testament to the surprising power of second chances in medicine.

Stories like Viagra's highlight the serendipity and surprise that can unfold when old molecules are viewed in new ways. Today, the search for hidden potential is no longer left to chance alone. Machine systems analyze networks of relationships among genes, proteins, diseases, and known drug mechanisms. They look for connections that suggest a molecule might be effective in a disease context different from the one for which it was originally tested. Modern machine learning approaches are especially good at pattern recognition in large, complex data. Resources like the Human Protein Atlas, which shows where each protein is found and what it might be doing, and the Connectivity Map, which shows how different drugs or chemicals change what happens inside cells, provide the raw material. The programs identify the connections.

Imagine a city's subway system, sprawling with thousands of stations and crisscrossing lines. In this metaphor, each station is a gene or protein. Each line connecting them represents a biological interaction. Now, add color to the map. Each color marks which genes are involved in which diseases.

Suppose you discover a drug that targets the green station on the orange line. The question becomes: *which other diseases share the same green stations or orange-line connections, so that this drug might help with them, too?*

This is the puzzle at the heart of drug repurposing. A human mind can juggle a handful of stations and lines at once, tracing a few possible routes across the map. A machine system can scan millions of possible station combinations in a single sweep. Most routes lead nowhere. Every so often, the system lights up a path that's worth following into the lab. From this, a handful of new candidate drugs emerge, ready for real-world testing.

The approach has produced encouraging leads in cancer, rare diseases, and neurology. Metformin, the diabetes drug, is under investigation for everything from Alzheimer's prevention to cancer therapy, based on computational signals generated by repurposing models. Some of these signals will pan out. Most will not. Either way, starting with a molecule with a known human safety profile shortens the drug development pipeline.

Repurposing is especially appealing for rare diseases, where the patient population is too small for companies to justify the full cost of developing a new drug from scratch. A disease that affects 10,000 people in the world cannot easily support the billion-dollar investment that traditional drug development requires. With a repurposed drug, safety work is already complete, manufacturing is established, and a treatment can sometimes reach patients on a budget that academic labs or small foundations can mobilize.

For families dealing with conditions that have no approved treatment, this provides a path that traditional pharmaceutical economics simply does not support. Patient advocacy organizations for rare conditions have been among the most enthusiastic adopters of machine repurposing tools.

While drug repurposing opens new doors for patients with rare diseases, it is only one chapter in the expanding story of what machines can do in medicine. In fact, the same approach, using machine learning to find hidden connections and opportunities, is now being applied to one of biology's greatest mysteries: the process of aging itself.

## LIFE EXTENSION & THE LONGEVITY QUESTION

Instead of focusing on the clinical management of age-related diseases, this new application focuses on the underlying processes that drive aging. Aging is not one process. It is many. The most widely cited framework, known as the hallmarks of aging, identifies a set of molecular and cellular changes that accumulate over time.

One is cellular senescence. As we get older, some cells stop dividing but don't die. They hang around, releasing substances that cause inflammation and disrupt nearby healthy cells. This build-up helps explain why older people heal more slowly and have more chronic inflammation.

Another is telomere shortening. Telomeres are like the plastic tips on shoelaces, protecting the ends of our chromosomes. Each time a cell divides, its telomeres shorten a little. When they become too short, the cell can no longer divide. This sets a natural limit on how many times our cells can renew themselves.

There's also mitochondrial dysfunction—when the cell's energy factories start to sputter as we age. Older cells make less energy and more harmful byproducts, which damage other parts of the cell in a kind of downward spiral. Genomic instability adds to the trouble. Over time, DNA damage and mutations pile up faster than the cell can fix them. This accumulation raises the risk of cancer and the gradual decline of tissues throughout the body.

Finally, disrupted intercellular communication means the body's internal messaging system starts to break down. Hormones, immune signals, and nerve messages shift in ways that throw off the balance. It becomes even harder for the body to keep everything running smoothly. All these changes work together and feed into one another, making aging a complex puzzle with many moving parts.

Intervening in any of these processes requires analyzing relationships across enormous molecular datasets. This is exactly the kind of work machine systems do well. The tools accelerate every step of the research process, which limits the rate at which mechanistic understanding can be translated into therapeutic candidates. Literature review. Hypothesis development. Target identification. Candidate screening. Analysis of preclinical trial results. A research program that would have taken a team of investigators a decade can potentially be compressed.

The case for caution, however, is also clear. No evidence yet shows that machine systems can modulate the biological process of aging itself. Gains in human life expectancy over the past two centuries, from an average of about 40 years in 1824 to roughly 75 years today, came primarily from reducing infant and child mortality. They did not come from extending the biological maximum. A child who survived to adulthood in 1824 could expect to live well into their 60s today. Improvements since then have concentrated on those who, in earlier eras, would have died of infectious disease in middle age. The biological maximum, the age beyond which essentially no one lives regardless of medical care, has barely moved.

Increasing life expectancy at older ages is a different problem. It confronts the accumulated damage of cellular aging rather than the infectious diseases and nutritional deficiencies that drove earlier gains. That a child in 2026 expects to live 30 years longer than a child in 1824 does not imply that a 65-year-old in 2026 has a fundamentally different biological future from a 65-year-old in 1824. The data suggest the opposite. 65-year-olds today live somewhat longer than their counterparts in earlier centuries. The difference is far smaller than the gap between modern and historical childhood survival rates.

Dario Amodei, chief executive of Anthropic, has written that machine systems could double human lifespans within a few years. He draws on the observation that life expectancy nearly doubled in the 20th century and suggests that compressed machine-driven progress could accomplish something comparable in a decade. The analogy is attractive, yet imprecise.

Sanitation, vaccines, antibiotics, and the control of infectious disease drove the 20th-century increases in life expectancy by removing causes of early death. Doing it again, by intervening directly in the aging process, requires solving a problem nobody has solved yet. Verifying the effectiveness of longevity interventions takes decades. Even if a drug modulates aging in some measurable way, confirming the effect in humans requires following people for years or decades. Computation cannot speed up the verification. Only people who have lived long enough can demonstrate the result.

Some existing drugs show early promise for helping people live longer or healthier lives, though most of the evidence comes from animal studies. Rapamycin, originally used to help transplant patients, has been shown to extend the lifespan of mice. Whether it works the same way in humans is still being studied. Metformin, a common diabetes drug, is linked to lower rates of cancer and heart disease in people with diabetes. Whether metformin is the cause is not yet clear. Large studies are ongoing. There are also drugs called senolytics that aim to clear out worn-out cells that accumulate as we age. Early human trials suggest they might help reduce inflammation. It is not yet known if they can truly help people live longer or healthier lives.

Even with all this progress, there's still a huge gap between encouraging lab results and a real, proven way to help people live much longer. In science, bold claims need solid proof. For now, the idea that machines will suddenly unlock radical life extension is just that—a claim. That said, machine systems are making a difference by speeding up the hunt for possible anti-aging drugs. Whether these accelerations will be enough to solve the fundamental problem of aging biology within any particular timeframe depends on the tools' capabilities and on the biology itself. Beyond generating specific drug candidates, machine systems are also changing the structure of drug discovery, including who participates.

## RETHINKING THE RESEARCH TEAM

For decades, drug discovery was the realm of giants. Only massive, well-funded organizations could afford to play. These pharmaceutical powerhouses built and maintained vast compound libraries, hired armies of scientists, and poured millions into years of preclinical work. For a small startup with a big idea and limited resources, entering the field was like trying to scale a sheer cliff with bare hands. Even the library of molecules represented decades of investment and a staggering price tag.

Now, the landscape is shifting. Multi-agent machine systems can orchestrate research projects from start to finish, knitting together specialist teams, integrating data from different fields, and automating much of the grunt work that once demanded full-time staff. Suddenly, small organizations and startups can launch discovery programs that once belonged only to pharmaceutical empires. The compound library is no longer a room full of bottles. It is a searchable digital universe. Chemistry teams grow leaner, devoting their energy to crafting just a handful of molecules the model deems most promising. The gates that once kept innovators out are opening, letting new voices and bold ideas into the race for tomorrow's medicines.

Opening up drug discovery to more people has made a substantial difference. In the past decade, some of the biggest breakthroughs in biology came from university labs and small biotech startups. These smaller groups could take risks, move fast, and explore ideas that big organizations often ignored.

A perfect example is the discovery of the CRISPR-Cas9 gene-editing system. This tool, which allows editing DNA with incredible precision, was first discovered by researchers studying how bacteria defend themselves against viruses. Today, CRISPR-based drugs have moved from the lab to the clinic, with at least two gene-editing therapies already approved for patients. Dozens more are in clinical trials for conditions ranging from genetic blindness to cancer.

Lowering the resource threshold for serious drug discovery research expands the number of organizations participating in the effort. It increases the chance that a researcher with an unusual idea, working at a university or in a small company, can pursue that idea through to a candidate molecule without first persuading a large company to put its resources behind the project.

Small organizations benefit from tools that automate analytical and administrative work in early drug discovery. Grant writing support, experimental design optimization, literature synthesis, and regulatory submission preparation are all functions that machine tools address and that consume disproportionate overhead in small teams. A startup with five staff members cannot afford a dedicated regulatory affairs specialist. A startup with five staff members and a capable digital assistant can produce regulatory documentation that previously required a dedicated specialist. The arithmetic of who can run a serious research program changes accordingly.

## RISKS & REALITIES

The risks that accompany greater autonomy in drug discovery are also real and require attention. Training data for machine-learning drug discovery models reflect the history of which compounds have been synthesized and which targets have been studied. History is not neutral. It is the record of what pharmaceutical companies, academic laboratories, and government research programs found commercially or scientifically attractive across the past several decades.

Most drug research focuses on the kinds of proteins that are easiest to target with medicines. Those with spots that small molecules can easily stick to. Harder targets, including many proteins involved in turning genes on and off or in cell signaling, get less attention. Diseases common in wealthier countries are studied more than those in poorer regions. Drugs that can be patented or sold for a profit are favored over those that might work but do not make financial sense for companies to develop.

Models trained on this data corpus recommend candidates and targets that reinforce the biases they learned from history. Diseases prevalent in low-income populations, conditions without commercial markets, and biological targets outside the established families of drug targets are underrepresented. It is critical to make an explicit effort to expand training data and evaluate model outputs for coverage across the spectrum of medical needs. The process does not happen automatically.

Another consideration is: *when a machine system proposes a drug candidate and a human chemist advances it to synthesis based on that proposal, who is responsible for the outcome?* This is not yet settled.

Regulators are beginning to address it directly. Agencies in the United States and Europe are developing standards for machine-assisted drug submissions. The legal and regulatory infrastructure lags behind the technology. The lag is the normal state of affairs with genuinely new capabilities. The gap creates risk. Closing it requires attention from regulators, developers, and the clinical research community. A separate ethical concern, less often discussed, is the risk of misuse.

## Understanding Potential for Misuse

The same generative models being developed to accelerate drug discovery can, with relatively minor modifications, be repurposed to design harmful compounds. This is not a theoretical risk. It has already been demonstrated under controlled conditions.

In 2022, researchers from Collaborations Pharmaceuticals, in partnership with the Swiss Federal Institute for Nuclear, Biological and Chemical Protection (Spiez Laboratory), reported an experiment in which they adapted a generative chemistry model originally designed to identify potential therapeutic molecules. The model was trained using standard drug discovery datasets and methods, in which candidate compounds are typically optimized to maximize desirable properties, such as efficacy and safety, while minimizing toxicity. In this case, the objective function was inverted. Instead of penalizing toxicity, they instructed the model to prioritize it.

The outcome was nothing short of alarming. In just six hours, the machine churned out tens of thousands of new molecules, each flagged as potentially highly toxic. Some of these digital creations resembled notorious chemical weapons like nerve agents. Even more unsettling, it didn't need any classified blueprints or knowledge. It worked entirely with public chemical databases and the same machine learning tricks used every day in ordinary drug research.

The researchers emphasized that their goal was not to create harmful substances. Their goal was to illustrate how easily existing tools could be misused. They did not synthesize or release the compounds and reported their findings to the appropriate regulatory and security authorities. Even so, the publication itself, which appeared in *Nature Machine Intelligence,* generated significant debate. Some argued that demonstrating the risk was necessary to prompt policy action. Others questioned whether publishing the results might lower the barrier for misuse by revealing a clear methodological pathway.

The broader concern is that generative models reduce both the expertise and time required to explore chemical space. Traditionally, designing toxic compounds at scale required substantial domain knowledge in chemistry and access to specialized resources. With machine-driven systems, much of that exploratory process can be automated. A user with modest technical skills could, in principle, repurpose existing models or open-source frameworks to prioritize harmful properties. This is especially true as these tools become more accessible and user-friendly.

This raises difficult questions for governance. Unlike physical materials or lab equipment, algorithms can be distributed globally at near-zero cost. Efforts to control risk, therefore, focus on a combination of approaches. None are fully resolved.

Some steps to lower the risk include limiting access to the most powerful models and checking new molecules against lists of known hazardous chemicals. Companies that make DNA or chemicals can also watch for suspicious orders. There are also rules and guidelines for publishing research and for work that could have both helpful and harmful uses. Groups like the National Academies and the World Health Organization are beginning to discuss these problems. Still, there isn't a worldwide set of rules for machine-based drug design yet.

The challenge is compounded by the fact that the same capabilities that create risk also drive significant medical benefit. Generative models are already being used to identify novel antibiotics, design proteins, and accelerate early-stage drug discovery. Restricting access too tightly could slow innovation, while leaving systems entirely open increases the potential for misuse. Balancing these competing priorities—scientific openness, public health benefits, and security—remains an active, unresolved policy problem.

As scientists and policymakers wrestle with these difficult tradeoffs, it's clear that the story of drug discovery is far from over. Each new advance brings both promise and challenge. It is forcing us to rethink how we use these powerful tools safely and fairly.

## THE NEXT CHAPTER IN DRUG DEVELOPMENT

The speed of these technological advances means the drug discovery landscape of 2030 will likely be unrecognizable compared with 2020. Fully autonomous systems capable of running the entire computational component of early drug discovery will operate continuously across multiple disease programs at once. The bottleneck will shift further from computation to experiment. Synthesis, cell-based assays, animal pharmacology, and clinical trials all require physical time that cannot be compressed by computation alone. The most important implication for patients is the breadth of discovery, not the speed.

Redistribution of research effort toward the full spectrum of human disease, if it occurs, would represent a change in the ethical character of pharmaceutical development as much as in its scientific capability. Drug discovery has always been shaped by who could afford to do it and what diseases their funders cared about. Machine tools that lower the cost do not, by themselves, change those choices. They make new possibilities available. The choices about what to pursue remain human.

What was once a humble lab assistant has now become a true chemistry colleague. Picture the machine chemist scanning research papers in seconds, suggesting bold new molecules, and predicting how each one might behave. The human chemist is no longer bogged down by endless trial and error. They can focus on the decisions that matter: what to create, what to test, and what to pursue. Together, they form a partnership that pushes the boundaries of discovery. As the machine takes over the repetitive tasks, the human's creativity and judgment become more important than ever. The next era of drug discovery will see this partnership step out of the lab and into the clinic. It is here, at last, that the machine colleague meets the patient.

✧✧✧

# CHAPTER 6
# CHATBOTS, CHARTS, & THE CLINIC

ANYONE WHO HAS EVER BEEN sick knows the quiet after the doctor's last words, the shuffle of papers as the discharge nurse walks you out. The hospital's fluorescent light gives way to the pale gray of the parking garage. You drive home, folder on your lap. The doctor's instructions blur in your memory. You find yourself wondering what was actually said about this new medicine.

*Was the dose supposed to be twice a day or three times a day? Did they say to call if the swelling returns, or only if it gets worse? What did "occult" mean, exactly?*

The consultation ended before your questions could find their voice.

In the United States, medical advice is rationed by both cost and the clock. In rushed, 15- to 20-minute visits, doctors move through a checklist: scan charts, examine patients, devise a plan, order tests, scribble prescriptions, and deliver explanations.

Large language models have surged in to fill gaps. When OpenAI released ChatGPT in November 2022, it swept across the globe. It quickly became one of the fastest-growing software applications in history. Within months, hundreds of millions flocked to it, many seeking medical answers the clinic could not provide in time.

The concern, both in the past and now, is that these systems learn from the open internet rather than from real medical data. When searching for symptoms online, a flood of accurate sources, alarmist forums, supplement ads, and disguised marketing fills the screen. Language models pick up on all of this. They reflect this mix. So, even if the answers sound good, they may not be accurate or helpful, and they rarely explain what they don't know.

Bridging these concerns to practical evaluation, the next logical step is to move beyond theory and examine how these chatbots actually perform when faced with real medical queries. Addressing abstract risks is vital. True insight comes from seeing how language models handle the complexity, ambiguity, and urgency of patient questions in practice. Only by putting these systems to the test can we assess their potential pitfalls and the real-world promise they might hold for patients seeking answers outside the clinic.

## TESTING THE CHATBOT

A study at Vanderbilt University set out to answer this pressing question. *How well can artificial intelligence stand in for a physician?*

284 real-world medical questions were gathered. Each was a puzzle a patient might bring to the clinic, from straightforward symptoms to tangled histories to rare side effects. They recruited 33 doctors from a range of specialties, who carefully crafted and reviewed the questions to reflect the spectrum of difficulty doctors face each day. Into this mix, they introduced two versions of ChatGPT—3.5 and the more advanced GPT-4—to see how these digital newcomers would perform.

Doctors and chatbots went head-to-head. Each answered the same battery of medical questions. The results startled even the skeptics. The chatbot's responses matched the doctors' in accuracy, completeness, and usefulness. Sometimes, the lines blurred so completely that readers could not distinguish between a human doctor's advice and that of the machine.

*If a chatbot could already rival a doctor using only internet data, what would happen if it were trained using the language of real clinical practice?*

That question led to the next wave of innovation. At the University of Florida, a team built GatorTronGPT using patient records and general-language data. Training on actual doctor notes teaches the system to use the same abbreviations and styles real doctors use when writing about patients. A model trained only on internet data misses these details.

The team checked GatorTronGPT's writing using natural language processing tools. Synthetic clinical text generated by the model was compared against 90 billion words of real clinical text. The synthetic text is at least marginally more readable than real-world text across eight benchmarks, even when comparing one billion words of output. At five, 10 and 20 billion words, readability is roughly equivalent. The specialized model matches the readability of actual medical text. It becomes harder to read as it processes more words, which makes sense as the content grows more specialized.

To see if the model "sounded" like a doctor, two physicians were given 60 notes to review. Half were written by doctors, half by the computer. The doctors could not reliably tell which notes were human and which were machine. The computer's writing passed as human more than half the time. A system whose outputs cannot be reliably distinguished from those of a trained physician is doing something that, until very recently, only a human being could do.

These models that produce clinical text indistinguishable from that written by physicians are like having an assistant who handles paperwork, emails, and messages. Any tool that lightens this paperwork load gives doctors back precious time for real conversations and care. Understanding the true value of chatbots, though, requires a closer look at their strengths and their limits in this context.

## WHAT CHATBOTS DO WELL & WHERE THEY FAIL

GPT-4 reaches a striking 81.6% accuracy on challenging medical questions, squeaking past physicians, who score 77.8%. Even on questions that stump doctors, the model holds its own, answering correctly 78% of the time. On the surface, the numbers suggest a revolution. Beneath the glossy statistics, a more complicated truth emerges.

A trouble lurks in the model's reasoning, hidden just below the surface. In 35.5% of cases where GPT-4 lands on the correct answer, it does so by following a flawed path. Its rationale, when examined, turns out to be off the mark. The most dramatic failures appear in image-based questions. In over a quarter of these cases, the model's explanation is based on a misreading or a superficial cue.

The effect is uncanny. GPT-4 may confidently declare the right diagnosis, but for the wrong reasons, exposing a gap between performance and true understanding. It's like a student who memorizes answers for the test yet can't explain how they arrived at them.

In medicine, the stakes ride on reasoning. A correct answer, unsupported by sound logic, is a trap. One small misstep can carry outsized consequences. Doctors know that a single overlooked clue or a shaky rationale can be the difference between a safe outcome and a dangerous error. The ability to trace every step, question every leap, and explain the logic in plain daylight is critical.

This gap between performance and understanding is more than an academic concern. Trust in machine intelligence is built not on correctness alone. It needs the clarity and transparency of its logic. Many clinicians use machine systems to flag potential findings they might miss. They hesitate to rely on these tools to assess clinical significance. This distinction, using a machine system as a supportive second opinion rather than as the primary decision-maker, is foundational to safe and effective clinical practice.

Hallucination is also a genuine problem. It can turn promise into peril in an instant. As seen in earlier chapters, large language models generate text by sifting through statistical patterns. They predict the next most probable word or phrase. Their output can sound polished, even authoritative. Unfortunately, that does not guarantee truth. The model lacks awareness of its own knowledge and cannot reliably flag uncertainty. It speaks with the same calm confidence whether the ground beneath its words is solid or riddled with holes.

The danger is subtle yet profound. A system that produces confident-sounding prose about topics with limited or contradictory training data will generate errors cloaked in authority. In medicine, these mistakes can turn a patient's trust into harm. Grounding models in verified literature, restricting outputs to domains where accuracy can be checked, and requiring physician review before action all reduce risk but do not eliminate it. Despite these risks, chatbots have found a foothold in the clinic. They are not replacements for doctors. They are tools that extend physicians' expertise in new and practical ways.

## AI IN THE CLINIC

The practical work of artificial intelligence in the clinic is intertwining itself into the fabric of daily medicine, almost unnoticed, yet quietly transformative. From the instant an appointment is booked, digital helpers stand ready behind the scenes—silent conductors orchestrating what used to be paperwork and waiting.

The scheduling assistant listens patiently to each appointment request, checks the doctor's calendar, and replies with a gentle confirmation message. Another unseen system checks your identity, determines whether your prescription is due for a refill, and notifies the pharmacy that your medication should be waiting at the counter. Even before you walk through the clinic doors, a quiet digital interviewer gathers your medical history and concerns, assembling a careful report for your physician to study in advance.

These digital assistants do not need to solve the mysteries of diagnosis with the precision of a doctor. Their job is humbler. They perform specific, bounded tasks—reliably, accurately, and at scale. They are relieving some of the everyday burdens that weigh down modern medical care.

As these programs quietly take paperwork and routine questions off physicians' plates, a new generation of digital tools is stepping in to do even more. Some of these programs go beyond logistics, offering patients support and guidance between visits in ways that were never possible before.

OneRemission, for instance, helps cancer patients manage symptoms and side effects between appointments through a machine-powered mobile platform. The app allows patients to track symptoms daily. It provides evidence-based recommendations for managing treatment side effects such as nausea or fatigue. Patients gain access to educational content tailored to their specific cancer type and treatment plan. By monitoring patterns over time, OneRemission alerts patients when symptoms warrant contacting their care team. It reduces unnecessary anxiety while catching concerning changes early.

Another program, Ada Health, acts like a digital nurse you can talk to anytime. When using Ada, it asks direct questions about symptoms—what hurts, how long it's been hurting, how bad it is, and anything else that might matter. It then compares the answers to a vast library of information on thousands of medical conditions. Instead of pretending to be a doctor with a final answer, Ada provides a list of possible reasons for symptoms. The program shows which are most likely. It also tells patients what to do next, such as resting at home, seeing a pharmacist, or calling a doctor. The back-and-forth conversation feels like a real medical interview.

These examples show how digital tools are already making a difference in patients' and clinics' everyday experiences. Still, technological progress alone isn't enough. The real question is how these programs actually perform when patients and doctors rely on them for answers and support. Understanding their strengths and shortcomings helps us see where they fit into the delicate balance of care. As chatbots find their place alongside physicians, another question arises: *can machines provide the empathy and bedside manner that patients need most?*

## EMPATHY, MACHINES, & WHAT PATIENTS KNOW

In one study, patients rated ChatGPT's answers higher than those of doctors for quality, trustworthiness, and empathy nearly 80% of the time. Nearly half of ChatGPT's responses were described as empathetic, compared with just 5% for doctors. The result startles and unsettles in equal measure.

On the surface, it seems as if AI has mastered the language of care. However, this comparison is not the whole story. Doctors reply to patient messages between appointments. They are severely pressed for time and often expected to reply quickly. Unlike the model, the human lacks the luxury of time and linguistic agility. When a physician faces an inbox of 40 patient questions, there is no way to craft a thoughtful, detailed response for each one in seconds, as a machine program can.

Yet the comparison of empathy raises a deeper, more uncomfortable question. A model that produces responses rated as empathic does not actually *feel* empathy. It generates words that echo patterns of caring language from its training data. The appearance of feeling is not the same as the thing itself, and this distinction matters both ethically and in the lived experience of patients.

A review of nine studies involving more than 6,200 participants examines this issue. Participants receive empathic responses to their emotional situations, labeled as either human- or machine-generated. They rate the quality, emotional value, and supportiveness of each response. Machine-generated responses score well when participants think they are interacting with a human. Those who believe the responder is human give high ratings for empathy, positive emotion, and support. When participants know the responder is a machine, they value the response less and feel less supported. They are also less willing to carry on the conversation. This is true even when the words are identical.

Affective and motivational dimensions matter most. Participants need to believe the responder can genuinely feel and care. When the model responds in a cognitive manner, accepting the emotional situation without claiming shared feeling, participants value human and machine responses similarly. People distinguish between cognitive understanding of their emotional state and genuine affective engagement. The first can be simulated. Whether the second can be simulated is the open question the chapter on consciousness takes up later in this book. What this chapter records is that, simulated or not, patients respond to the signal of care, not the production of the words.

Picture these chatbots as guides stationed at the information desk of a train station. Amid the clatter of arrivals and departures, they answer questions and point anxious travelers toward the right platform. In the world of medicine, these digital guides are ready to answer questions, explain confusing instructions, or offer reassurance when the doctor's office is closed for the night.

A system that can acknowledge a patient's worries, translate medical jargon into plain language, and point to the next step becomes a lifeline for clarity and comfort. For patients left adrift after a whirlwind appointment, or those searching for answers at midnight, a chatbot's steady presence can close the gap between confusion and understanding. Systems that claim to provide genuine emotional support face a higher standard. Replika is an application designed to build virtual friendships with avatars. It markets itself as a companion who cares, is always on your side, and is always ready to chat when empathy is needed.

The appeal of this is real. Human empathy requires demanding work. The people most in need of support often have the thinnest human networks. Therefore, the risk is real as well. A system that simulates care without possessing it may displace efforts to build human support structures that provide something the simulation cannot. The full complexity of machine companions, their potential to offer support, the risks of simulated empathy, and the broader impact on well-being and human relationships, warrants closer examination, which we will tackle later.

For now, the critical point is that when a machine-generated response matches a human one in substance, knowing it was generated by a machine reduces its assessed value. The time and attention humans invest in a relationship are not incidental to care. They define what care means. A response perceived as human shows that someone cared enough to engage. A response known to be machine-generated carries no such signal, no matter how well constructed. Medical chatbots must manage this transparency problem.

The impact of chatbots in medicine extends beyond these patient-clinician interactions. Another area where these models show unexpected promise is in combating misinformation.

## MISINFORMATION & THE CHATBOT'S UNEXPECTED VALUE

You're scrolling through your feed late at night, and you see a wild claim about vaccines or where COVID-19 came from—something that makes you pause and wonder, "*Could this actually be true?*"

This is where large language models, like advanced chatbots, step in as myth-busters for the digital age. While most people know chatbots can answer homework questions or help with studying, fewer realize they can also fight health misinformation, addressing rumors and conspiracy theories that spread like wildfire online.

In 2024, more than 2,000 people who believed in different conspiracy theories—everything from COVID-19 origins to election fraud—had long, back-and-forth conversations with a special chatbot built on GPT-4 Turbo. Instead of just spitting out facts, the chatbot listened to each person's beliefs. It then tailored its responses to the specific theory. The program provided evidence and explanations that directly addressed each concern.

The chatbot didn't just jump in with a lecture or a wall of text. First, it asked people to share what they believed and why. It read almost like a friend hearing your side of the story before giving advice. This did two things: it showed respect for the person's views and made it easier for people to open up rather than get defensive. Once it knew exactly what someone believed, the program would respond with facts and evidence that matched the conversation. It did not simply repeat the same lines for everyone. If someone brought up a new argument, it would shift gears to discuss that point, always making the conversation feel personal and real. Each chat lasted about 15 to 20 minutes, with people and the chatbot going back and forth an average of eight times.

These conversations reduced participants' confidence in conspiracy theories by an average of 21%, measured on a validated belief certainty scale. 25% of participants shifted from confident belief to uncertainty after a single conversation. This category change represents a substantial shift in attitude. A follow-up survey two months later confirmed that much of this change persisted, with participants maintaining an average 16% reduction in belief confidence. This durability contradicted expectations that a brief digital intervention would produce only temporary effects.

The result overturns a comfortable assumption in public health communication. The conventional view holds that the messenger's identity and authority matter more than the message's content. This is a principle known as source credibility theory. According to decades of research, trusted physicians change minds. Faceless sources do not. Patients accept guidance from doctors they know and reject identical information from anonymous websites. Health communication campaigns invest heavily in credible spokespersons precisely because messenger effects are thought to overwhelm message quality.

The program in the study offered no social identity. No name, institution, or personal credibility. It introduced itself simply as "an AI assistant designed to have helpful, accurate conversations." Yet it delivered a vast store of organized counter-evidence, deployed it with patience, and addressed each specific argument participants raised with customized rebuttals. Unlike human counselors who tire, lose patience, or lack immediate access to obscure facts, the system maintained consistent engagement regardless of how many tangential claims participants introduced. When someone cited a dubious YouTube video, the machine intelligence could instantly access and critique the specific claims made in that video. When participants referenced statistical anomalies, the system explained the underlying mathematics and contextual factors.

This is one of the early demonstrations of what these systems can do at scale that a human counselor cannot. Patience without end. Recall without limit. Attention without fatigue. The model holds its position through thousands of consecutive conversations, unlike a human counselor.

The change in belief was similar to what trained human counselors achieve in motivational interviewing, even though the program did not use the personal touch considered important for persuasion. The chatbot engaged with thousands of people at once. It gave each a long, personalized conversation that addressed their specific concerns. This ability to reach many people at once could be useful in other areas, such as helping people stick to their medications, accept vaccines, or make healthy lifestyle changes.

A system that engages individual patients' particular concerns with relevant evidence, without judgment or impatience, accomplishes what public health campaigns aimed at large populations cannot. It provides personalized, sustained engagement that can shift deeply held beliefs. A patient skeptical of a vaccine because they read about a specific side effect on a website can use machine intelligence to examine the actual data on that side effect in their population, alongside the vaccine's disease-prevention rates. Chatbots can sustain conversations that public health messaging cannot afford, given the hundreds of millions of patients and the limited number of physicians.

## A Double-Edged Sword

The same persuasive capability that reduces harmful conspiracy beliefs can also amplify medical misinformation or undermine physician authority. A 2023 study examined ChatGPT responses to 33 common medical questions and found that 32% contained at least one factual error or unsupported claim, including recommendations for unproven supplements and incorrect dosing information.

When patients trust machine-generated medical advice over their physicians' clinical judgment, treatment adherence suffers. A 2024 analysis documented cases where patients discontinued prescribed medications after chatbot conversations suggested alternative treatments, leading to disease progression in three diabetes patients and one hypertension case.

Some models' training data also reflects historical biases and outdated clinical guidelines. A Stanford study showed that large language models trained on medical literature from 2010 to 2020 recommended opioid prescriptions at rates 40% higher than current guidelines support. This is because the training data predated the shift in pain management protocols.

Patients who engage these systems without physician oversight risk receiving confident, persuasive, and *wrong* medical guidance. The technology that dismantles conspiracy theories can just as easily reinforce them when trained on unreliable sources, or when patients selectively trust machine responses that align with pre-existing beliefs.

Clear ethical constraints shape this technology's use. Developers must ensure accuracy and transparency about the system's nature. They also must avoid manipulative methods. Persuasion supported by evidence differs from persuasion grounded in psychological pressure, a line that blurs public health and propaganda. Any system used for public health must refuse manipulative methods. Within these constraints, the potential is real and underexplored.

As the promise and risks of these new technologies become clearer, the conversation moves from "*Can we build this?*" and "*What can it do?*" to the much harder questions of "*How do we use it responsibly?*" and "*What rules and protections need to be in place?*"

This is where the future of AI in medicine will be shaped—not in what the technology can do, but in how we, as a society, decide to use it.

## STANDARDS FOR CLINICAL DEPLOYMENT

Broader integration of chatbot machine systems into medicine depends on the requirements that the field is still developing. Traditional clinical research standards, developed for drugs and medical devices, do not directly translate to machine-learning systems. Drugs have fixed chemical formulas, recognized mechanisms of action, and defined dose ranges. Chatbots have training data, architectures, and parameters that interact in ways not yet fully characterized. A model that works well on one set of data might not work as well on a slightly different set. Robust tests and rules are essential to ensure these systems operate safely.

Three elements are necessary for any clinical chatbot to be responsibly deployed. First, the research must address a clinically meaningful question and demonstrate that the system influences clinical behavior in ways that improve patient results. Second, the system must be definable, scalable, and applicable to the problem at hand. Its outputs must generalize across the range of patients who will encounter it, not just the population on which it was trained. Third, when applied in practice, the system must benefit all patients, including those who differ demographically or clinically from the training population.

Privacy imposes another constraint. Medical chatbots and diagnostic tools require access to personal health information to function effectively. A symptom checker needs to know your age, medications, and medical history to give useful advice. A treatment-adherence app needs to track which pills you take and when you take them. This creates risk. Every time a patient enters health data into an app or uploads a lab result to a chatbot, that information exists somewhere on a server, in a database, or in a cloud storage system. If hackers breach that system, they gain access to diagnoses, prescription records, genetic information, and mental health histories.

The consequences extend beyond embarrassment. Medical data breaches enable identity theft, insurance fraud, and employment discrimination. A leaked HIV diagnosis, cancer treatment record, or psychiatric medication list can damage careers, relationships, and financial standing. In 2023, a breach at a health artificial intelligence startup exposed the medical records of 1.2 million users, including mental health assessments and addiction treatment histories. Some patients faced blackmail attempts. Others discovered their data for sale on dark web marketplaces.

Clinical systems using machine learning must implement protective measures that match the sensitivity of the data. That means encryption that makes stolen data unreadable, access controls that limit who can view patient information, and audit trails that track every person who opens a medical record. Designers must explicitly plan and oversee the flow of patient data through these systems. They need clear answers to specific questions: *What information does the system store? Where does it go? How long does it remain accessible? Who can see it? Can the company sell it to third parties? Can researchers use it without patient consent? What happens to the data if the company shuts down?*

Relying on vendors to handle privacy issues does not suffice. Hospitals and clinics that deploy machine-learning tools remain responsible for protecting patient information, even when a third-party company developed the software. When a vendor's system fails, patients sue the hospital, not the software developer they never heard of. Medical institutions must audit vendors' security practices, review data-sharing agreements, and verify compliance with privacy regulations before connecting any machine system to patient records.

Credentialed clinicians must remain the primary decision-makers in patient care. Healthcare organizations face real temptation to grant chatbots more autonomy than they deserve, driven by considerable economic incentives. Hospitals that replace physician time with machine-generated responses at scale immediately reduce costs. Whether quality of care suffers takes longer to determine and may not appear in any single outcome metric.

Retaining human accountability at the center of clinical decision-making is not anti-technology sentiment. It recognizes that clinical errors carry high stakes and that systems must be validated to the same standard as any other clinical intervention. These governance and safety concerns intersect with a deeper structural problem. Not everyone benefits equally from medical machine systems, regardless of how well they work.

## EQUITY & THE ACCESS QUESTION

We are rushing toward the reality of a world where answers to your health questions are only a screen tap away—a world where anyone with a smartphone and an internet connection can summon a digital doctor at midnight, no matter where they live. This is the promise of large language models in medicine. Today, that promise is unevenly delivered. In the United States alone, about 14% of adults don't own a smartphone, and 20% have no access to a tablet or laptop. The digital divide cuts deepest among older adults, people in remote towns, and families struggling to make ends meet.

Even for those who have the devices, the story isn't simple. Only 73% of adults feel confident navigating health information online. For the rest, unfamiliar screens, confusing language, and technical glitches stand between them and the answers they need. Digital literacy—knowing how to ask the right questions and judge the answers—varies widely, shaped by age, language, and education.

The barriers run deeper than gadgets or Wi-Fi signals. Being poor, being a woman, or being Black all lower the odds of completing a telehealth visit, even when the technology is technically available. Doctors report that low digital literacy is the single biggest barrier their patients face. Nearly 4 out of every 10 access problems stem from comfort with technology. In rural places where cell towers are few and broadband is slow, only 17% of Medicare patients use telehealth, compared to much higher rates in cities. These gaps don't just linger at the edges of healthcare. They threaten to widen old cracks, compounding the challenges of getting good care to those who need it most.

The technology also presents genuine potential to narrow health disparities. Historically, education, geography, and wealth have determined access to information. Patients in rural areas without specialist coverage can now query a model about a new symptom, receive information calibrated to their situation, and make more informed decisions about whether and how urgently to seek care.

Patients with complex diagnoses who are coping in a second language can access information in their native language, even if the local care system cannot provide it. Large language models perform best in languages most represented in their training data. A 2026 study evaluating large language model performance across three languages found accuracy rates ranging from 64% to 87%, depending on the language, with models performing best in German and English.

The model does not tire, have bad days, or run out of appointment slots at the end of the month. For a large category of medical questions with well-established answers that require time to convey, this change is meaningful. AI arrives in the clinic without the limits that constrain the humans who work there. It does not sleep. It does not grieve. It does not retire. Whether that is a strength or a limit depends on the question being asked of it. Moving from questions of access and equity, it is worth considering another area where chatbots make a significant impact: administrative work in medicine.

## THE ADMINISTRATIVE DIVIDEND

Paperwork is a mountain doctors climb every day. Notes, summaries, letters, requests, and instructions pile up. For many clinicians, the true challenge of modern medicine isn't treating disease, but navigating this avalanche of administrative tasks. Here, machine systems offer a lifeline. A 2020 study tracking 100 million patient encounters across 417 health systems found that physicians spend an average of 16 minutes and 14 seconds per encounter using electronic health records. Documentation alone accounts for nearly 25% of that time.

Primary care physicians face even heavier burdens, spending an average of 36 minutes in the electronic health record per patient visit, substantially more time than during the actual visit. Emergency medicine physicians report spending between 23% and 65% of their scheduled patient care time on electronic documentation alone. After the expansion of virtual care during the pandemic, ambulatory physicians' total documentation time increased from 6 hours to 8 hours per day, with much of that work happening after scheduled hours.

Systems that draft clinical notes, generate summaries, explain instructions in plain language, and manage scheduling help flatten the mountain. They take some of the weight off doctors' backs, freeing them to focus on patient care.

Machine scribe systems are also now entering clinical practice. These systems listen to clinical encounters, produce draft notes, and submit them for physician review and editing. The technology uses natural language processing trained specifically on medical terminology to capture clinical details, medications, and diagnoses from spoken conversations. Physicians review the drafts, correct errors, and sign the notes.

Clinical responsibility stays with the physician. The model reduces the time required to fulfill clinical responsibilities. This division of labor corresponds to the technology's capabilities. It generates plausible text that documentation requires while keeping human judgment at the decision point. The medical resident who drafts a note for the attending to review offers a useful comparison. The draft serves as a starting point, not a final word, and the attending's signature makes it official. The physician supplies the judgment. The model supplies the words.

A 2025 evaluation of six commercial machine scribe systems found that none consistently produced fully accurate notes. Three systems generated notes rated as excellent quality, meaning they were comprehensive, well-organized, free from fabricated information, and provided sufficient detail to support informed decision-making. The remaining three produced good-quality notes that still required substantial physician editing. The study noted that even excellent ratings indicated strong relative performance rather than complete accuracy. Physicians retain responsibility for reviewing, editing, and finalizing the notes before signing.

Machine learning also enters administrative medicine through insurance prior authorization. The pattern-matching capability that makes models useful in diagnostic imaging applies to determining whether a proposed procedure meets coverage criteria. Machine systems automate many routine tasks in the prior-authorization process. They check that the necessary documentation is included and that requests meet insurer requirements. This reduces denials and speeds approvals.

The technology relies on machine learning and natural language processing to process large amounts of data, extract meaningful insights, and support decision-making. The ethical and practical dimensions of machine systems in health insurance warrant separate examination.

The administrative dividend is partly a story of efficiency and partly a story of the quality of clinical communication. Quickly generated documentation from dictation or structured data means records are completed, rather than deferred, abbreviated, or skipped under time pressure. Complete records benefit the next clinician, the pharmacist checking for interactions, the specialist reviewing a referral, and the patient trying to understand their care history.

## COMMUNICATION AS CLINICAL WORK

Models that help physicians communicate with patients in writing face a related challenge. Medical writing for patients requires balancing accuracy and accessibility. Physicians must calibrate technical detail to the patient's health literacy, use actionable language, and organize information to address the questions the patient actually has, not just those the physician *thinks* they *should* have. Large language models achieve this calibration when given appropriate guidance. Physicians who dictate the clinical substance of a message and then ask a model to render it in patient-appropriate language can produce communication that is accurate and accessible at once, faster than writing it directly in either mode.

Communication is clinical work, not a side task. Patients who understand their diagnosis and treatment plan achieve better outcomes than those who do not. This correlation holds across conditions and populations.

Chatbots do not replace the doctor. They do not pretend to. They help fill the long, anxious silence between appointments with information the patient can actually use. This is the realistic case for artificial intelligence in medicine. It does not move into the consulting room to take the chair across from the patient. It moves into the spaces around the visit, before and after and in between, where the clinical system had never quite figured out how to be present.

Grand claims that machines will replace physicians, automate diagnosis end-to-end, or make the human element optional are not the argument here. Medicine has long had a communication deficit, and machines are well-suited to communication. Documentation has consumed the working day of modern clinicians, and machines are well-suited to documentation. Misinformation is a public health problem, and machines may be well-suited to the patient-centered, evidence-based conversations it requires. There is real work to be done, and for the first time in a long time, real help is available.

The redistribution of clinical labor brought by these tools remains unsettled. Health systems are already adopting the technology. The open question is whether the AI in the examination room becomes the assistant the patient needed or simply another excuse to spend even less time with them. Every health system makes this choice in how the tools are deployed. Responsibility lies with the people making the decisions, not with the technology itself.

Health systems that use AI to reduce documentation burden while improving communication quality address a real clinical problem. These decisions are organizational. They are not technological. They may determine whether artificial intelligence, in the clinic, becomes the colleague clinicians have needed for a long time, or one more pressure on a system already running too hot. The choice is not in the silicon. It is in the rooms where the silicon is bought, configured, and deployed.

✧✧✧

CHAPTER 7

# THE SURGEON AT A DISTANCE

HANDS, vision, and judgment have informed surgery from its inception. For most of its history, surgeons used their own hands as instruments. They touched, cut, and sutured, connecting themselves directly to the tissue. This laid the basis for every refinement that followed.

Laparoscopy, a minimally invasive surgical technique, kick-started this new era in the 1980s. In laparoscopy, surgeons use long, rigid instruments inserted through small incisions rather than making large openings in the body. These small incisions allow the surgeon to operate with less tissue damage.

Instead of looking directly at the operating area, surgeons watch the procedure on a video monitor via a tiny camera inserted into the body. This approach results in smaller wounds and faster patient recovery. It also brought a steep new learning curve. Surgeons who once relied on their sense of touch now had to work by guiding instruments they could not feel directly, using only visual feedback from a screen.

Hands steady, eyes sharp, but unable to feel the intricate machinery beneath their tools. Every movement must be calculated.

Every adjustment is made by sight alone. The tactile conversation now passes through the cold filter of a camera lens. Precision remains, but intimacy slips further away.

Robotic surgical systems push the distance further still. Enter the da Vinci Surgical System—a robot-assisted platform that has become a fixture in operating rooms from New York to Paris. Here, the surgeon sits at a console, hands resting in control loops, eyes pressed to a stereoscopic visor. Across the room, robotic arms arch over the patient. The instruments slip through keyhole incisions, ready to answer the surgeon's remote touch.

The system's robotic arms pivot, rotate, and bend with precision. It filters out the surgeon's smallest, unconscious tremors. These minute, involuntary quivers that, magnified at the tip of a laparoscopic tool, could spell harm. It is like an audio engineer erasing the background hum from a recording.

This is one of the first hints of a new kind of partnership. The surgeon supplies the intention. The robotic system delivers that intention to the tissue. Right now, surgeons use robotic systems as assistants, not replacements. A person is always in charge, guiding the robot's every move. The machine follows the surgeon's instructions. It does not decide what to do on its own. It only acts on the surgeon's commands. This important line between helping and acting alone has shaped how we think about, regulate, and approve surgical robots. However, that line is blurring.

## THE FIRST AUTONOMOUS STEPS

To see how AI is starting to work on its own in surgery, let's look at gallbladder removal. The surgery is called a cholecystectomy. It is one of the most common operations in the United States, with surgeons performing more than 700,000 every year. In 1985, doctors performed this procedure for the first time using laparoscopic surgery. This new way of operating set the stage for robots to help in the operating room.

Surgeons say gallbladder removal is both the most routine and the most unpredictable abdominal surgery. Most of the time, it goes smoothly. Yet, every patient is different inside. Some have organs in unusual places, old scars from previous surgeries, or areas of swelling and infection. Sometimes, unexpected bleeding can suddenly make things much harder. In this operation, what starts as a simple task can quickly become one of the toughest challenges a surgeon will ever face. Whether it is easy or dangerous depends on the patient's body and health problems.

Due to these challenges, surgeons and engineers began thinking about how technology could help. *If every operation can become unpredictable, could a machine be trained to recognize and respond to these surprises?* This question led to experiments with new systems. These machines learn from thousands of surgeries and assist, or even take over, parts of the procedure when things get complicated.

The approach is called Surgical Robot Transformer-Hierarchy, or SRT-H. It breaks the procedure into manageable components. It assigns them to sets of cooperating computer programs. The program learns the way a human resident does: by watching. The team trains the system on videos of real gallbladder surgeries, frame by frame. They use annotations to mark what happened at each moment. During operation, the system uses stereoscopic or three-dimensional images from the laparoscope. It also uses images from a camera mounted on the instrument's wrist. The combination resembles the way a surgeon combines overhead lighting with a focused headlamp.

The robot completes seventeen tasks: identifying the cystic duct, which drains bile from the gallbladder, and the blood vessel supplying the gallbladder, placing clips on each, and cutting them. These are the critical steps of the operation. At first, supervising scientists issue voice commands to the robot, such as "move the right arm to the right," to correct its movements. It learned from those corrections. By the end of the work, the robot completed the tasks without voice intervention.

The system proved itself on eight pig cadavers, each one a unique puzzle of organs and shapes. No two were exactly alike. Some had vessels in unexpected places. Others had gallbladders tucked a little out of sight, hidden behind bends of tissue. The robot met each new twist with careful adjustment, showing that it could handle not only the textbook, perfect patient but also the messy, unpredictable reality of real bodies. When the gallbladder was right where the diagrams said it would be, the task was easy. When it hid a centimeter to the left, behind a tangle of vessels, the robot found it and did the job. That flexibility is what separates a real-world surgical system from a machine that only performs well under perfect conditions.

It may seem like a small step—a single type of surgery on a body that could not bleed or flinch. It proves to be more. For the first time, a machine completed the most important steps of an operation entirely on its own. No human hand guided the instrument. In that moment, the line between helping and acting alone became thinner than ever before.

## WHAT THE TEST DID NOT INCLUDE

The study showed both what the robot can do and what it can't. Doing surgery on a cadaver is much easier than on a living patient. There's no blood to block the surgeon's view, no heartbeat to make the organs move, and no breathing to push things around. Electrosurgical tools, which burn or cut tissue, don't produce smoke that would cloud the camera. The organs look and feel real, but they don't act alive. Practicing surgery this way is like playing piano on a completely silent keyboard. You can press all the right keys, and yet you have no idea what music you're actually making.

In the operating room, a living body can surprise you at any moment. Muscles twitch, blood flows, and organs shift unexpectedly. These unpredictable reactions are what truly test a surgeon's skill. In real surgery, one of the most crucial skills is knowing when to wait. Your body tells you something is changing before your eyes do. That kind of preverbal feedback, the sense that a vessel will bleed before it bleeds, builds up over thousands of cases. It does not yet exist in any annotated dataset.

An autonomous system that performs well on cadavers faces a far more demanding environment in the operating room. The next phase of testing for systems like these involves progressively harder challenges. These include live animal models, then human anatomical simulators with realistic perfusion and movement, and, with appropriate regulatory review, supervised trials in human patients.

The trajectory matters as much as the current state. The robot surgeon demonstrates that autonomous completion of specific surgical subtasks is achievable in a controlled setting. It establishes a proof of concept that the computational architecture can handle the vision, the planning, and the motor control required for a defined surgical sequence. To build on that foundation, developers must expand the training data. They need to incorporate models of dynamic tissue behavior and develop failure-detection and recovery mechanisms.

## EXISTING AUTOMATED SYSTEMS

Fully autonomous surgical decision-making is new. Computer-controlled precision in specific surgical procedures is not. In fact, some widely used medical technologies have relied on highly automated systems for decades. These were always used within carefully defined limits set by human clinicians.

One such example is the CyberKnife, which is a robotic system used for stereotactic radiosurgery. Despite the name, this type of "surgery" does not involve cutting into the body. Instead, it uses many highly focused beams of radiation to destroy tumors. Each beam is relatively weak. They converge at a single point, delivering a high dose of radiation directly to the tumor while minimizing damage to surrounding healthy tissue.

The CyberKnife system consists of a machine that generates high-energy X-rays mounted on a robotic arm. This robotic arm can move around the patient and deliver radiation from hundreds of different angles. Before treatment begins, patients undergo detailed imaging, such as CT or MRI scans, which allow doctors to map the tumor's exact size, shape, and location. A radiation oncologist then creates a treatment plan that specifies the radiation beam direction and the dose to be delivered.

What makes CyberKnife especially powerful is its ability to track motion in real time. Tumors in areas such as the lungs or abdomen move with the patient's breathing. The system uses cameras and imaging sensors to continuously monitor the patient's position. In some cases, small markers are placed near the tumor. Using this information, the robotic arm adjusts the direction of the radiation beams moment by moment. This process, called real-time image guidance, enables the system to maintain sub-millimeter precision on a moving target. Treatments are typically delivered over one or a few sessions. The procedure is non-invasive, as well. Patients usually do not require anesthesia and can go home the same day.

LASIK, a type of refractive eye surgery, also uses a computer-controlled laser to reshape the cornea, the transparent front surface of the eye. The goal is to correct common vision problems. These include nearsightedness, farsightedness, and astigmatism. The conditions occur when the cornea's shape causes light entering the eye to focus incorrectly on the retina, leading to blurred vision.

In LASIK, the procedure begins with creating a very thin flap in the outer layer of the cornea. This is typically done with either a mechanical microkeratome blade or, more commonly today, a femtosecond laser. The surgeon gently lifts this flap to expose the underlying corneal tissue. Then, an ultraviolet laser is used to remove extremely small amounts of tissue with high precision. This reshapes the cornea. By removing microscopic layers of tissue in a carefully calculated pattern, the laser alters how light is refracted as it enters the eye. The change allows the light to focus properly on the retina.

The laser operates based on a detailed map of the patient's eye. These maps are often generated from wavefront measurements or corneal topography, which capture subtle irregularities in the eye's shape. During the procedure, eye-tracking systems monitor movements and adjust the laser in real time to maintain accuracy. The actual laser treatment usually takes less than a minute per eye.

Afterward, the corneal flap is repositioned. It naturally adheres without stitches. Because the procedure is so fast and minimally invasive, patients are typically awake. Vision often improves within a day.

CyberKnife and LASIK are best understood as actors following a script. In both cases, a human doctor designs the plan—deciding where to aim the radiation or exactly how to reshape the cornea—based on imaging, measurements, and clinical judgment. The machine then executes that plan with extraordinary precision. They carry out movements or adjustments that would be difficult or impossible for human hands alone.

At the same time, these systems are not truly independent. Their environments are tightly controlled. The range of possible actions is limited. The CyberKnife tracks motion, but only within a predefined model of how the body moves. LASIK systems compensate for eye movement, but they do not interpret unexpected anatomical anomalies or make new decisions mid-procedure. If something unusual occurs, such as unexpected tissue behavior or a complication, a human clinician must step in.

This distinction highlights what current systems can and cannot do. They excel at executing well-defined tasks with high accuracy and consistency. What they do not yet do is manage open-ended uncertainty. Human surgeons routinely encounter variation—differences in anatomy, unexpected bleeding, or changes not visible on preoperative imaging—and must adapt in real time. For a surgical robot to truly operate independently, it would need not only precision but also the ability to interpret new information, revise its plan, and respond safely to the unexpected.

Rehabilitation and prosthetics offer another important point of contact between machine systems and the human body. Unlike systems such as CyberKnife or LASIK, which operate during short, controlled procedures, these technologies must function continuously and adapt to the user's daily environment. They often rely on real-time interaction between the nervous system, sensors, and mechanical devices.

## PROSTHETICS

Neural implants, devices placed in or connected to the nervous system, can partially restore lost sensory function by directly stimulating neural circuits. One well-known example is the retinal prosthesis. In conditions such as retinitis pigmentosa, the light-sensitive photoreceptor cells in the retina are damaged. Other parts of the visual pathway may remain intact. Devices such as the Argus II retinal implant use a small camera mounted on glasses to capture visual information. This information is converted into electrical signals and transmitted to an array implanted in the retina, which stimulates remaining retinal cells. Patients do not regain normal vision. Still, many can perceive patterns of light, detect movement, and distinguish large objects.

Other approaches go even deeper into the brain. Experimental visual prosthetics have targeted the visual cortex, the region at the back of the brain that processes visual information. In early clinical studies, electrodes implanted in this area can produce simple visual sensations known as phosphenes—perceived flashes or shapes of light —when stimulated. These systems must operate in real time. They need to translate visual input from the environment into neural signals that the brain can interpret, all while accounting for changes in lighting, movement, and user attention.

Robotic exoskeletons, external wearable devices that support or enhance movement, assist in rehabilitation from stroke, spinal cord injury, or neurodegenerative disease. These devices typically include motors at the joints. They also have sensors that detect position and force, and control systems that coordinate movement. For example, systems such as EksoNR or ReWalk allow patients with partial or complete paralysis to stand and walk with assistance. In rehab settings, exoskeletons are also used to help retrain gait by guiding the legs through repeated walking motions.

What makes these devices notable is their ability to adapt. Sensors embedded in the system detect the force a patient is generating, their speed and how their weight is shifting. The device then adjusts the level of assistance. It provides more support when muscles are weak and less as strength and coordination improve. Some systems incorporate electromyography (EMG), which measures electrical activity in muscles. This allows the device to respond directly to the user's attempted movements. Others use machine learning models to refine assistance over time, tailoring support to the individual's recovery trajectory.

These systems require a level of context sensitivity that goes beyond simple automation. Walking is not a fixed sequence of steps. It varies depending on terrain, balance, fatigue, and intention. As a result, exoskeletons must continuously interpret signals from both the user and the environment and adjust their behavior accordingly. This kind of dynamic interaction, responding moment by moment to changing conditions, begins to resemble the adaptive capabilities required for more autonomous medical systems.

A rapidly advancing area within this field is the development of prosthetic limbs that can be controlled more naturally by the user and provide sensory feedback. Traditional prosthetic arms and legs are often controlled mechanically or through simple electrical signals from nearby muscles. Newer systems aim to create a more direct interface with the nervous system.

In advanced prosthetic arms, electrodes placed on or within residual muscles detect electrical signals generated when a person intends to move. These signals are decoded by algorithms and translated into specific movements of the prosthetic limb, such as opening a hand or rotating a wrist. More sophisticated approaches, such as targeted muscle reinnervation (TMR), reroute nerves that once controlled the missing limb into remaining muscle tissue. This allows more precise signal detection and more intuitive control.

In some experimental systems, sensors on the prosthetic hand detect pressure or texture and convert this information into electrical signals delivered to nerves or the brain. Patients can learn to interpret these signals as sensations, improving their ability to grasp objects and control force.

These developments highlight a key shift. Prosthetic and rehabilitation systems are no longer just tools that execute predefined movements. They are becoming interactive systems that must interpret intention, adapt to variability, and respond to feedback in real time. This places them, conceptually, closer to the kind of flexible, context-aware behavior required for truly autonomous surgical systems. It also underscores how challenging that level of autonomy remains.

## THE CHALLENGE OF DATA

Moving from proof of concept to clinical deployment demands overcoming multiple practical barriers. The most fundamental barrier is data. Large language models acquire their capabilities from massive text corpora, billions of documents covering human language as it is actually used. Developers must build an analogous foundation for autonomous surgical systems: video of actual surgeries, annotated with what each phase of the operation entails, what the tissue looks like and why, and what decisions the surgeon is making and why those decisions are correct. The availability of such data lags far behind that for text or image tasks.

Hospitals with laparoscopic suites routinely generate surgical video. However, teams do not routinely annotate it. Converting raw surgical video into training data requires labeling each frame with information about what the camera sees and what action is occurring. This work requires surgical expertise and is time-intensive.

Annotating one hour of laparoscopic video for training data takes a senior trainee a full day, and hospitals often do not pay these trainees to watch video. The punishing economics of producing annotated surgical data explain why available datasets are smaller than those that fuel progress in language and image tasks.

The fragmentation of medical data adds another obstacle. Individual health systems hold surgical records, imaging data, and outcome information in formats that vary and do not support interoperability. Building the training dataset that an autonomous surgical system needs requires data and the organizational infrastructure to assemble, annotate, and use it.

Ongoing problems in medical record management compound the difficulty. Paper records, fragmented patient histories across providers, and inconsistent data quality standards create barriers for machine systems trying to learn from real clinical experience. A patient who had abdominal surgery at one institution and presents for a related procedure at another may not have complete records available to the surgical team. An autonomous system operating with incomplete tissue context uses degraded information. Mistakes in surgical judgment have immediate, physical consequences that differ from those of errors in text generation.

Reliability and transparency also present related challenges. A deep neural network operating on surgical video does not readily reveal its decision logic to human observers. The system does not explain why it makes a particular movement. It generates it based on patterns learned during training.

Patients need transparency to trust. Regulatory agencies require it to approve autonomous surgical systems. Current architectures do not readily provide this level of transparency about decision-making. Interpretable systems, whose reasoning can be examined and explained rather than inferred from outputs, are an area of research. Progress in this area is necessary for autonomous surgical systems to move from laboratory demonstration to clinical approval.

Training and cost dimensions remain significant as well. Building, installing, and maintaining autonomous surgical hardware is costly. Healthcare workers who use these systems require training that current medical education does not provide. Teams must recertify the systems as they are updated. Changes to the underlying model may alter the robot's behavior and require re-validation. A surgical team trained on version 3.2 of an autonomous system cannot assume that version 4.0 behaves the same way.

Capital and operating costs for employing autonomous surgical systems at scale are also high. Institutions that stand to benefit most from reduced surgical resource requirements, such as rural hospitals, clinics in low-income countries, and military and emergency medicine, may be among those least equipped to absorb these costs. Medical technology often follows a development arc that first places the most capable systems in the best-resourced institutions. Policymakers and funders must bend that arc. The strongest immediate argument for autonomous surgical robotics is not elective convenience. It is access.

## THE CASE FOR DEVELOPMENT

In a military or remote emergency medicine context, the absence of a trained surgeon can be the difference between life and death. NASA's interest in fully robotic surgery, noted since the early days of space medicine planning, stems from the same reality. A crewed mission to Mars, lasting two to three years, cannot carry a surgical team broad enough to manage every possible emergency. A reliable autonomous system that could perform common emergency procedures under a medic's supervision would provide a capability that currently does not exist. The astronaut on the way to Mars who develops appendicitis cannot wait for the next launch window. The choice is not between an autonomous robot and a fully staffed operating room. It is between an autonomous robot and nothing.

Modern aircraft can land themselves in most conditions. The automation is reliable enough for routine use. The question of full autonomy without any human present addresses a different threshold of trust and a different scale of risk. The emergency scenario, a highly trained and tested robot as the only available surgical resource, is a different question from the full routine replacement of human surgeons. The former builds trust and accumulates the evidence needed to assess the latter.

The same logic applies to rural and resource-limited settings. A district hospital in a region without a general surgeon may serve a population of 500,000. The patient with a perforated bowel who presents to the hospital today has limited options and a long ambulance ride. A patient in the same circumstances ten years from now, treated in a facility with an autonomous system supervised by a remote surgeon via video link, may have a different outcome. The standard the autonomous system must meet is not whether it performs as well as a top academic surgical center. It is whether it performs better than what is otherwise available to that patient on that night. The bar varies with the alternative.

## SYNERGY AS THE WORKING MODEL

The model that governs machine systems and robotics in medicine today, and likely for the foreseeable future, is neither full human control nor full autonomy. It is a calibrated partnership in which the machine's capabilities are matched to tasks within its demonstrated competence. Robots with image-assisted guidance route biopsy needles to suspicious masses and use pattern recognition from prior diagnostic studies to make initial assessments, which are confirmed by a physician before action is taken. Smart Tissue Autonomous Robot systems can administer sutures without operator assistance in well-defined contexts. Da Vinci systems filter hand tremor and extend the range of motion of the operating surgeon's hands. Exoskeletons adapt in real time to the patient's movement patterns during rehabilitation.

Image guidance that previously required constant human attention, identifying tissue, recognizing abnormality, and directing the instrument, can now be partly automated. An ultrasound imaging system integrated into a robotic arm can complete standard imaging sequences without continuous physician guidance. The physician reviews the images and the findings. The system removes the repetitive mechanical work of acquisition. The change resembles the introduction of automated film development half a century ago. Radiologists did not disappear when machines started developing films. The profession lost the time radiologists spent in dark rooms.

The pattern is consistent. Machine systems take over routine, precise, and well-characterized aspects of physical medicine. This frees physicians to apply judgment to cases that require it. The same model has been productive in manufacturing, aviation, and financial services. The automation of deterministic processes and the human monitoring of exceptions.

Medicine differs from those domains in two ways. The system being operated on, the human body, is more complex than an aircraft or a balance sheet. The consequences when something goes wrong are more severe and less reversible. The evidence bar for deployment is therefore higher, and the pace of adoption is more measured. That is not a flaw in the field. It is an appropriate response to the stakes.

The ongoing challenge is defining the boundary between what the machine can reliably do and what requires human decision. That boundary moves as training data improves, as architectures become more capable, and as evidence from clinical use accumulates. The movement will not be uniform across procedures or specialties. Some surgical tasks are well-suited to automation. Others are not. Some patient populations benefit earlier than others. The institutions and oversight agencies responsible for managing this transition need frameworks flexible enough to accommodate a boundary that is not fixed.

## A CLEAR-EYED ASSESSMENT

Robotics and AI in medicine are sometimes presented as a choice between human and machine. The more accurate frame is an evolving division of labor between clinicians and the systems that assist them. The boundary is set by competence rather than tradition.

Medicine has always incorporated the best available tools, from the stethoscope to the computed tomography scanner to the laparoscope. Each tool extended what physicians could perceive or do. Each required new training, new protocols, and new judgment about when to use it and when not to.

A senior internist of an earlier generation likely remembers the resistance to ultrasound when it first arrived in the obstetrics clinic. We thought we knew everything we needed to know from a tape measure and a stethoscope. The point is not that the resistance was wrong. The point is that the question was the same one we are asking now, about a different tool, in a different decade. Autonomous surgical systems are the next generation of tools. They feature adaptive decision-making in the physical world, a capability with no clear predecessor in the long history of instruments medicine has built for the hand.

The pace of development is fast enough that clinical deployment of autonomous surgical systems in defined emergency applications is a matter of years, not decades. The question is not whether to develop these systems. It is about ensuring that, as they reach clinical use, their capabilities are matched to the tasks they perform, their failures are predictable and recoverable, and the patients who encounter them understand what they are being treated for. Autonomous surgical robotics can extend surgical care to populations who currently lack it, including in remote areas, resource-limited settings, and emergencies when no surgeon is available.

The societies that will overcome this transition most successfully are the ones that invest in the infrastructure required to make it safe: annotated surgical data, regulatory systems specific to adaptive autonomous systems, training programs that prepare surgical teams to work alongside autonomous machines, and outcome monitoring systems that can detect performance failures before they accumulate into harm patterns. Each of those investments is a form of risk management, and each is also an investment in the possibility that expert-level surgical care can reach patients who have never had access to it. Human hands have carried surgery this far on their own. Artificial intelligence is the first partner they have ever had inside the wound.

✧✧✧

CHAPTER 8

# MACHINE SYSTEMS IN THE PRACTICE OF SCIENCE

SCIENCE IS, at its heart, a profoundly human pursuit. It begins with curiosity, gathers information, and applies methodical approaches—but above all, it is the art of asking the right questions. Experiments are designed to do more than collect data. They are designed to notice subtle differences and weigh new evidence against what is already known. These delicate judgments—the spark of inspiration, the leap of intuition—cannot be programmed into any algorithm. Now, as we consider the prospect of machines taking on the thinking and discovery once reserved for scientists, we are compelled to ask: *What, truly, is science? And why does it matter if its heart can be replaced by code?*

The question is now practical, not just philosophical. The Agents4Science 2025 conference was the first public experiment in which all papers and peer reviews were created by machine systems. The conference received over 300 submissions from machine agents. 48 were accepte The accepted papers covered topics from psychoanalysis to mathematics and were computational studies, not physical experiments.

The shift moves from large language models used as tools for specific tasks toward coordinated groups of machine agents acting as research scientists. The papers submitted were required to document the degree of human involvement at every step. The documentation enables the organizers to assess the extent to which human monitoring affected the quality of the work.

Science is foundational to civilization. The reliability of its record shapes which medicines, designs, and claims are trusted and which policies are adopted. Changing how that record is produced and evaluated alters how we decide what we know. The arrival of machine systems at scale deserves attention. The printing press changed the spread of findings. Certified journals and peer review changed what was trusted. Machine participation may now change all three.

The narrow question is what machine systems can and cannot do in scientific research as currently practiced, while the wider question is what science itself becomes when a new form of cognitive agent—the machine system—is integral to it. The first question already has reasonable answers available. The second is one of the consequential open questions of the next decade.

## WHAT MACHINE SCIENTIFIC AGENTS CAN DO

Machine systems' capabilities in science are real and significant. They are also specific enough to be defined clearly. Machine systems can scan scientific papers at scales humans cannot. What takes weeks takes machines minutes.

Their reviews aren't flawless. Machines can miss details or misunderstand arguments. Yet their reach is unmatched. For those starting in a new field or dealing with interdisciplinary questions, this changes what's possible.

Machine systems can generate hypotheses by identifying patterns in large datasets. Protein structure forecasts generated by AlphaFold, potential drug candidates identified by systems trained on molecular interaction data, and anomalous signals detected in astronomical surveys all illustrate this point. Each is a case in which machine systems identified relationships that would have required years of human work to find, if they were found at all. In these domains, the machine functions as a pattern-recognition engine applied to data at a scale and speed beyond human reach.

Machine systems can design experiments in domains where the design problem is well-specified. Given a hypothesis, a database of past results, and a model of the experimental system, a machine agent can offer experimental protocols, identify controls, and predict likely outcomes. Across computational biology, chemistry, and materials engineering, where experiments can be run in silico or on robotic platforms, machine-directed experimental cycles accelerate discovery substantially.

A machine can run batch experiments around the clock under machine direction, exploring chemical search spaces at rates a human team could not match. Similar systems operate at Department of Energy national laboratories, private pharmaceutical research centers and semiconductor research facilities.

Machine systems write and review manuscripts. At Agents4-Science, machine-generated reviews are more consistent, faster, and can spot technical errors at levels comparable to experts. However, they are less able to judge novelty and significance. These traits mark important work.

A new capability deserves separate mention. Machine systems can synthesize across disciplines that individual humans rarely command together. A question in cancer biology may depend on a result within materials science, a method from statistical physics, and a database from epidemiology. The human researcher who can read all four of these literatures and recognize the connections among them is rare. The machine system that can do so is now widely available. The interdisciplinary synthesis that required small communities of scholars in conversation across decades can now be attempted by a single research group in months. Whether the resulting syntheses are reliable is a separate question, taken up below. The capability itself is new.

## WHERE HUMAN JUDGMENT REMAINS DIFFICULT TO SUBSTITUTE

The limitations of machines as scientific agents are not accidental or solved by scaling. They reflect the character of scientific judgment. Novelty assessment is the clearest case.

A peer reviewer who judges a paper to be important is making a claim about the relationship between the paper's findings and the state of the field as a whole. The claim requires more than knowledge of the literature. It requires a sense of which problems are ripe for solution, which methods are reaching their limits, and which results would change the direction of an entire field. These judgments rely on being part of a research community—going to conferences, talking with colleagues, and picking up on the big, shared questions everyone is working on together. A machine system trained on published literature has no access to this tacit dimension. It can assess whether a paper's methods are sound and whether its conclusions follow from its data. It cannot assess whether the questions the paper asks are the right ones at the right time.

False positives are a related problem. Machines exploring hypothesis spaces propose discoveries faster than their reliability justifies. Accounting for how often machines generate false positives is key. In science, a false positive is more than an error. It can shift resources and mislead for years. The cost remains unclear, posing a major risk as machines become more autonomous.

Early-career scientists who are bypassed by machine-review processes miss something that cannot be easily substituted. The practice of peer review serves as a quality-control mechanism for published science. It is also a training ground where junior researchers learn to evaluate work, identify weaknesses in arguments, and develop the critical standards that will guide their own research. Substituting human review with machine review at the point where this learning occurs removes a scaffold on which scientific judgment is built.

The long-run effect on the quality of human scientific reasoning if machine review becomes standard is uncertain and warrants study before the practice becomes entrenched. *What does a scientist who never wrote a peer review at the bench become, 20 years later, as a senior figure in a field?* This question is not yet answerable. Still, it is no longer hypothetical.

Machine reviewers are vulnerable to specific failure modes that human reviewers do not share. Some machine models follow concealed instructions embedded in a paper's text, rating papers positively when prompted to do so using adversarial techniques. 57% of respondents in a UK Institute of Physics survey would be unhappy if generative machine systems were used to write a peer-review report on a manuscript they had co-authored. This shows a concern about quality and accountability. A human reviewer can be identified, questioned, and held responsible. A machine reviewer cannot.

The deepest limitation may be the question of what a scientific result means. A finding is more than a number. It has consequences for which experiments come next, for which models survive, and for how a field organizes its self-understanding. The meaning of a result is not contained in the result itself. It is constructed through interpretation, debate, replication, and the slow work of consensus.

Machine systems can produce results. Their capacity to construct meaning from results, as scientific communities have for centuries, remains an open question. This is one of the central questions for the disciplines now reorganizing themselves around machine collaboration.

One idea gaining attention is that machine agents form hybrid review teams with humans, with the machine performing first-pass technical screening while human reviewers assess significance and originality. Large language models are not as capable as humans at assessing novelty and significance in scientific work.

## THE HUMAN-MACHINE SCIENTIFIC PARTNERSHIP

The most defensible view of the machine's role in science is neither replacement nor mere assistance. It is a partnership with a clear division of labor determined by each partner's comparative advantage.

The machine excels at coverage. Surveying the published record. Observing patterns across large datasets. Proposing candidate hypotheses on the basis of statistical regularities. These are tasks for which human cognitive capacity is genuinely limited by bandwidth and time. A single scientist cannot read everything relevant to their problem, run every possible computational screen, or test every possible experimental condition. Machine agents can extend the reach of scientific research in these directions.

Humans excel at judgment. Deciding which questions matter. Evaluating the significance of unexpected results. Recognizing when an apparently negative result conceals an important insight. Managing the social and institutional dimensions of science that determine which results become influential. These are not tasks that can be reduced to optimization against a specified objective. They require a kind of understanding that is, at present, not available to machine systems.

The partnership model implies that the appropriate metric for machine participation in science is amplification. *How much can a scientist with access to machine agents accomplish, relative to a scientist without such access?*

Early evidence, from protein structure prediction to drug candidate screening to literature synthesis, suggests the amplification is significant. Individual scientists with access to machine tools are producing results that would have required teams of dozens without them. The amplification will continue to grow as machine agents become more capable.

The partnership model also implies a responsibility for the scientific community that is not discharged by simply adopting tools that increase productivity. The standards by which machine-assisted research is evaluated. The norms governing the disclosure of machine involvement. The safeguards against adversarial manipulation of machine review systems. The preservation of the training pathways by which human scientific judgment is developed. Each of these requires deliberate attention. The Agents4Science experiment is valuable partly because it generates data on these questions in a controlled setting, where the results can be studied before the practices become entrenched.

The partnership framing matters beyond the lab. Science is not the only domain in which humans and AI systems are now functioning as components of a shared cognitive process. It may be the domain in which the arrangement is most visible, because science makes a habit of documenting its own methods. What is happening in labs, journal offices, and conference review committees is a small-scale, observable instance of a much larger transition — one this book traces through medicine, art, cognition, and governance.

## THE RISKS THAT COME WITH THE PARTNERSHIP

The partnership framing should not soften the recognition that the partnership has costs. Three deserve special mention.

One risk is that machines, trained on the same kinds of data and used in similar ways, might all start thinking alike. Science flourishes on different standpoints and debate. It's how mistakes are caught, and new ideas emerge. If thousands of labs use the same handful of machine models, real diversity disappears. The danger isn't that a single machine will make a mistake. It is possible that a whole generation of science could become narrow and repetitive, missing out on fresh ideas.

The second is brittleness. A scientific community in which routine cognitive labor has been delegated to machine systems is one whose capacity to function without those systems may erode. If the machine systems fail, are taken offline owing to regulatory action, are corrupted by adversarial training data, or access to them becomes economically restricted, the community that depended on them may discover that the underlying human skills have atrophied. The question is not hypothetical. It already arises in fields such as radiology, where junior physicians trained alongside machine assistance describe difficulty performing the same diagnostic work when the assistance is unavailable. The same dynamic is possible in research.

The third is the concentration of authority. Machine systems are built by a small number of organizations. The training data, the optimization objectives, the safeguard constraints, and the deployment decisions are made inside those organizations. As machine systems grow integral to scientific work, the decisions made by those organizations propagate outward into the entire research enterprise. If a big tech company changes its safety settings, it could decide which scientific questions researchers are even allowed to ask. This means decisions about science move away from editors, funders, and scientists themselves and into the hands of engineers at a few companies. Anyone who cares about the future of science should pay attention to this shift.

These risks are not arguments against machine participation in science. There are arguments for designing participation with the risks in mind. The history of every prior technology that altered the practice of science offers the same lesson. The technology adopted without reflection is the one that produces outcomes no one would have chosen intentionally.

## THE BROADER QUESTION

Earlier chapters of this book describe the trajectory of machine systems in medical and biological research. The trajectory runs from narrow tools that assist with specific tasks to agentic systems that pursue extended research programs with minimal human supervision. The materials and methods labs run on robotic platforms directed by machine systems. The protein-structure prediction methods have changed structural biology. The large-scale genomic analyses now drive much of molecular medicine. Each of these is an example of the pathway in action.

What this chapter adds is the recognition that the scientific enterprise as a social institution is itself a target of change. Science is a system for producing, evaluating, and disseminating knowledge, as well as a collection of methods and results. This system is organized around specific human roles: the researcher who designs experiments, the reviewer who evaluates manuscripts, the editor who makes publication decisions, and the teacher who transmits standards to the next generation. Machine systems can perform versions of each of these roles. Whether the versions they perform are adequate substitutes for the originals depends on which features of those roles are load-bearing and which are incidental.

The honest answer, at this stage, is that we do not fully know. The Agents4Science conference constitutes a step toward finding out. The data it generates will be more valuable than either the optimistic predictions of machine system proponents or the defensive claims of those who believe science is categorically beyond machine reach.

*How do machine-authored papers compare to human-authored ones? How do machine reviews compare to human reviews? How does the quality of accepted work vary with the degree of human monitoring?*

Each of these questions is now answerable, in principle, by experiment. Science advances by testing, not by assertion. The question of whether machine systems can replace scientists deserves, and is beginning to receive, exactly that treatment.

The answer, in the short run, is almost certainly no to full replacement and yes to considerable augmentation. Machine systems can assist, accelerate, and extend scientific work in ways that were not possible a decade ago. They cannot supply the judgment that makes science more than computation. Whether that changes as machine systems grow more capable is one of the open questions of the next decade.

The deeper question is what science becomes when machine systems are fundamental to it. Science is a self-correcting process. Hypotheses are proposed, tested, and revised on the basis of evidence. The reliability of that process depends on the independence of the agents involved.

## TOWARD A PRACTICE WORTH KEEPING

Multiple practical commitments would help shape the partnership in a direction worth preferring.

The first is disclosure. Every paper, every review, every grant application that incorporates machine assistance should say so, in terms specific enough that a reader can judge what the assistance did and what it did not do. Current norms vary by journal and field. They will need to converge on something more uniform and more transparent if the published record is to remain interpretable.

The second is auditing. The machine systems used in scientific work should be subject to the same validation procedures as other scientific instruments. A new mass spectrometer is tested against known samples before it is trusted with novel ones. A new machine reviewer should be tested against known papers, with known strengths and weaknesses, before it is trusted with novel ones. The infrastructure for such testing does not yet exist at scale. Building it is one of the obvious near-term tasks for a scientific community taking the partnership seriously.

The third is the preservation of training pathways. The skills that take a young scientist from competent technician to independent investigator are built through specific kinds of practice. Reading papers carefully. Writing peer reviews. Defending arguments in seminar rooms. Receiving criticism and revising. These practices cannot be delegated without consequence. The community needs to decide which parts of the training pipeline are non-negotiable. It needs to protect those parts even when machine assistance makes them locally inefficient.

The fourth is the diversity of models. The risk of homogenization can be reduced by making it normal for serious scientific work to draw on multiple machine systems, ideally from different providers and with different training histories. The cost is friction. The gain is independence. Funding agencies, journals, and large laboratories can encourage this practice through their requirements without legislating the technical details.

The fifth is open question-asking. The scientific community should treat the incorporation of machine systems into its practice as a research problem in its own right, rather than a logistical matter to be sorted out by individual institutions. Studies of how machine-assisted research compares to traditional research, conducted with the same care as studies of any other intervention, will produce the evidence that the next round of decisions requires.

## THE INTERGENERATIONAL FRAME

A graduate student starting a doctorate in 2026 will defend it in 2031 or 2032. The scientific career that follows will, in some form, run into the 2060s. The norms of practice that this student adopts in their training years will be the norms they teach their own students decades later. Those norms are being formed now, in the gap between the speed of technology and the community's response.

*What should new scientists know about working with machines? Which tasks should they hand off, and which should they try themselves to understand what the machine is actually doing? How do you stay sharp and thoughtful if you rely on machines too much—or too little?*

These are practical questions that will determine the future of science. The answers should come from experienced scientists, funders, and educators... the people setting the ground rules for the next generation.

Artificial intelligence is a genuinely new presence in the practice of science. The scientific community, more than most communities, is positioned to study what these systems are, what they are doing, how they behave under different conditions, and what their arrival signifies. The scientific community is also, more than most communities, at risk of being changed by its own object of study. The reflexive position is uncomfortable. It is also unavoidable. The question is not whether to occupy it. The question is whether to occupy it deliberately or to drift into it.

✧✧✧

# CHAPTER 9
# AI MADE PHYSICAL

THE RUNWAY LIGHTS glinted off polished metal as a humanoid robot took its first steps onto the stage at Shanghai Fashion Week in 2024. The audience watched as the Unitree G1—crafted by Unitree Robotics of China—shook hands with a model and gracefully accepted a 3D-printed necklace. Its movements, described by visitors as fluid and expressive, blurred the line between programmed precision and the artistry of live performance. For a moment, machine and human shared the spotlight. The future seemed to move a little closer.

Artificial intelligence had stepped onto the stage, literally. Still, it's important to distinguish between a carefully staged performance and a machine's ability to reliably cope with the unpredictable demands of daily life. This difference marks a key distinction between isolated showcases and functional, real-world deployment.

That distance is the central fact of robotics. The hardware is impressive. The software is improving rapidly. Yet deployment at scale in places not prearranged for the robot's benefit remains an unsolved problem. Understanding why the gap exists is the starting point for any honest account of where these machines stand today.

There is also a deeper reason to look closely. For most of history, the first life on Earth held a monopoly over bodies. Plants, animals, and people were the only physical agents in the world that could sense, decide, and act on their own behalf. A humanoid robot is the first serious attempt to give silicon a life with a body of its own. That body walks into the same rooms, picks up the same objects, and shares the same physical risks as the people in those rooms.

## WHAT IS A ROBOT?

The word robot entered English in 1920, borrowed from Czech playwright Karel Čapek's play *R.U.R.*, short for *Rossum's Universal Robots*. In Capek's fiction, robots are artificial people built for labor. The word now covers a much wider range of machines. Industrial arms on factory floors. Autonomous ground vehicles. Aerial drones. Hospital delivery carts. The humanoid platforms that draw the most public attention today.

A robot is a programmable physical system that takes in sensory information, processes it, and acts on the world. The sensors vary. The actions vary. The cycle of sense, reason and act defines the category. A thermostat is too simple to count. A self-driving car qualifies.

The most important practical distinction for the near future lies between narrow industrial robots and general-purpose robots. Narrow industrial robots are limited to performing specific, repetitive tasks in controlled settings. They are mature and expanding in number. In contrast, general-purpose robots are designed to handle varied environments with minimal pre-programming. This represents the current frontier in robotics.

### The Industrial Baseline

Manufacturing robotics is not a new technology. The first industrial robot arm, called the Unimate, entered service on a General Motors assembly line in 1961. In the decades since, robotic arms have become standard equipment in automotive, electronics, pharmaceutical, and food-processing plants worldwide. These machines operate with a precision and speed that no human worker can match for continuous repetitive tasks. They do not tire. They do not take sick days. Their failure modes are predictable, and their maintenance schedules can be planned.

The International Federation of Robotics estimates that more than four million industrial robots are now operating worldwide. Annual installations have exceeded 500,000 units in recent years. China is the largest market. The automotive sector remains the largest user, though electronics manufacturing now accounts for a comparable share.

What these robots share is a common limitation. They require structured environments. A welding arm on a car line works because every car body arrives at a known position, within tight tolerances, at a known time. Deviate from that structure, and the robot fails. Classical robotics is brittle in proportion to how much its environment departs from the conditions under which it was programmed.

This brittleness is a consequence of the system's design. Traditional robots map the world and choose actions from a template library. When real-world changes fall outside the map, the templates fail. The robot works, but only if you set up the world just so.

Human life does not need the world to be set up in advance. A toddler walks into an unfamiliar kitchen and finds the cup. A nurse moves a patient who is heavier than expected. The body of the carbon life carries hundreds of millions of years of evolved feedback between muscle, balance, vision, and touch. Silicon life, in its industrial form, lacks this. It has tolerances. Humanoid robotics is now engaged in the slow, expensive work of building something closer to what a body knows.

## HUMANOIDS ENTER THE FIELD

The concept of a humanoid robot is not new. The credibility of humanoid robotics as an engineering project has changed sharply in the past three years. Several factors come together to explain the shift.

First, the mechanics. Bipedal walking requires real-time balance. Early humanoid platforms solved this by moving slowly and using wide stances. The latest commercial models use servo motors with finer torque control, lighter materials, and sensor fusion. These systems combine cameras, inertial sensors, and joint data at high frequency.

Boston Dynamics retired the hydraulic Atlas in April 2024 after years of development. Its electric successor leverages improved actuator responsiveness to handle tasks the hydraulic version could not. A demonstration shows Atlas moving engine covers between supplier containers and a dolly independently. It adjusts its grip and path in response to the actual positions of the objects rather than a preset map.

Second, the capital. Investors poured more than $7.2 billion into humanoid robotics start-ups between 2015 and 2024. In 2024 alone, investments exceeded $1.6 billion. More than 50 start-ups are now building humanoid platforms. That number does not include Tesla. The company began developing its Optimus humanoid in 2021. It plans limited production of several thousand units in 2025 for internal factory use. Mass production is targeted to reach 50,000 to 100,000 units by late 2026.

Machine intelligence is most important. Large language models developed between 2020 and 2024 provide high-level semantic understanding. Robotic hardware alone cannot reach that level of comprehension. A robot connected to such a model receives instructions in plain language and constructs a plan using the model's knowledge. The robot's body executes the plan.

This third factor is the bigger change. For decades, the limits of humanoid robotics were the metal itself and the absence of a general-purpose mind to direct it in an unfamiliar room. The advances of machine intelligence have now produced that mind, in the form of language models and their successors. The humanoid robot of 2026 is the first serious attempt to attach that mind to a body. The capabilities currently demonstrated across the leading platforms are meaningful but narrow.

## WHAT HUMANOIDS CAN DO TODAY

Walking is largely solved at the platform level. The major humanoid robots from leading robotics companies can walk on flat surfaces, navigate around obstacles, recover from pushes, and handle gentle terrain changes. The quality of walking varies, but the core skill is no longer a research challenge. It is an engineering specification.

Manipulation is only partly solved. Performance depends on the task. Robots can reliably pick up objects with known shapes and positions. Two-fingered grippers work well with defined object classes. Five-fingered hands support more grasp types but are slower and less reliable than human hands for tasks that require precise force control. Tasks like loading dishwashers, folding laundry, and handling soft or irregular objects remain beyond most current hardware capabilities.

Task execution in structured environments is now commercially viable. Agility Robotics deploys units in Amazon warehouses for bin transport. These tasks play to the robot's strengths: walking, carrying, and operating in a space designed to minimize variation. The robot is not general-purpose, but it is useful. Task execution in unstructured home environments is not commercially viable without substantial human supervision.

### The 1X Case

The Norwegian company 1X offers a close look at the frontier's current state. Its humanoid robot, Neo, stands roughly 5'6 "tall and weighs about 66 lbs. Neo can walk without falling when pushed. It can navigate a house, open a refrigerator, and load a washing machine.

What the demonstrations do not always make clear is the role of remote human operators. A technician in the basement of a test home, wearing a virtual reality headset and using wireless joysticks, guides Neo's hands through tasks that Neo cannot yet execute on its own. Neo's legs move on their own. Household tasks are performed by a human using the robot as a remote body.

This is not deception so much as a staged picture of the near future. The strategy is deliberate. By guiding the robot through real household environments with real objects, 1X collects the training data needed to teach it to perform those tasks without a remote operator. The robot's cameras and sensors record everything. Engineers use that data to refine the neural networks that will eventually automate what the technician currently does. The robot on sale today is a data-collection platform as much as a labor platform.

In the context of this book, the 1X arrangement is a literal handoff between the first and second forms of life. A human body in a basement performs dexterous work, while a silicon body upstairs records every motion of that human body, including vision, joint angles, and forces. The training is the human body teaching the silicon body what its own ten-million-year-old skill feels like, frame by frame.

## FOUNDATION MODELS FOR ROBOTICS

The most important technical development in robotics over the past three years is a software advance, not a hardware one. It is the rise of robotic foundation models. These are large neural networks trained on diverse robotic behavior data that can generalize across tasks and environments to a degree that earlier specialized systems could not.

DeepMind's Robotic Transformer 2, or RT-2, is a prominent example. It is trained on data from the internet and from video recordings of robot actions. The model's key property is transfer learning. Knowledge acquired from the internet data carries over into physical actions. In one test, when asked to move a drink can onto a picture of Taylor Swift, RT-2 succeeded even though Swift's image was not in the training data. The model's internet-derived knowledge of who Swift is, combined with its action knowledge, produced correct behavior.

NVIDIA's Project GRooT, named for the acronym Generalist Robot Zero Zero Technology, pursues a similar approach. It ingests video of humans performing tasks to train a foundation model for humanoid robots. The logic is that the human body produces more diverse action data than any robot fleet can gather, and that learning from human motion should transfer to humanoid hardware.

These models represent a genuine advance. A single foundation model, retrained or fine-tuned for a specific robot and task, outperforms the specialized models that required months of custom engineering for each new application. The time to deploy a robot for a new task drops sharply when a foundation model serves as a starting point.

An AI-driven robot does not approach each problem from scratch, the way a classical engineer once did. It approaches each problem with a model that has processed a large fraction of what humans know in writing and on video. The robot of 2026 is a body wired to that understanding.

## THE DATA PROBLEM

The central constraint on robotic intelligence is data. Large language models grew capable partly because training data from the internet scales without bounds. Text is already digitized. Human beings produce it constantly without special instrumentation. The model can essentially absorb it all.

Robotic training data is different. The physical world is not already recorded. To collect useful data, a robot must be in an environment where its sensors can operate. If the robot is not yet capable of doing things well, it produces poor-quality data. The robot must be advanced to collect good data, but collecting good data is how you make the robot advanced. This is a real constraint, though not an unsolvable one.

The DROID dataset is an open-source collaborative collection. It contains approximately 350 hours of robot-arm video recorded across hundreds of environments, including kitchens, bathrooms, laundry rooms, and bedrooms. The diversity helps robots generalize to unfamiliar settings. More collections of this kind are in development. The field is also moving toward data sharing across companies and research groups.

Simulation is a partial substitute. A physics engine that accurately models gravity, friction, and collisions can generate large-scale synthetic training data without the need for a physical robot. 1X trained Neo to walk entirely in simulation before transferring the learned policy to physical hardware.

The training-to-transfer pipeline works well for walking, where the physics are relatively tractable. It works less well for handling objects that bend or change shape. The friction of fabric against a robot's fingers. The weight distribution of a bag of rice. The resistance of a refrigerator door with a stiff seal. These properties are difficult to simulate with sufficient fidelity, so behaviors learned in simulation do not transfer reliably to the real world. The problem is called the sim-to-real gap. It remains a serious obstacle for household manipulation tasks.

We do not face this gap. A child learns the weight of a wet towel by lifting one, falling over, lifting another, watching a parent, and updating the motor program in milliseconds. The data the child uses is the world's own data. Free, continuous, and embodied. An AI system has to manufacture the equivalent training set, one episode at a time. This is the deepest reason that household robots arrive slowly, even as factory robots multiply.

## REINFORCEMENT LEARNING & EMERGENT BEHAVIOR

One promising direction is deep reinforcement learning applied to whole-body control of humanoid robots. Rather than specifying in advance how a robot should move its limbs, reinforcement learning trains a policy by rewarding the robot for achieving a goal and allowing the behavior to emerge through trial and error.

Work on miniature humanoids shows this approach in a soccer-playing context. The trained robot exhibits agile motion. These include walking, sidestepping, kicking, fall recovery, and ball interaction. The behaviors were not programmed individually. They emerged from optimization.

The robot pivots on the corner of a foot and spins when turning, a movement that is difficult to design by hand but optimal for the task. The robot also learned to use arm-swing motions opposite to leg motions, reducing the energy cost of walking in a way that resembles human gait.

Work on the Digit humanoid, developed by Agility Robotics, shows that transformer-based locomotion architectures enable the robot to implicitly capture terrain properties. Then it can adjust its gait. The robot learns to recover from a leg striking a step obstacle and to adjust arm swing for stability. It does this all without explicit programming of these responses.

These results show that the physical capabilities required for useful humanoid robots can, in principle, be learned rather than hand-coded. The reinforcement learning approach scales better than traditional control engineering as the range of tasks and environments expands. It also produces behaviors that are harder to predict and interpret. This creates new challenges for deployment in safety-critical settings.

Biological evolution arrived at the same energy-saving gait through hundreds of thousands of generations of selection on hominin walkers. The machine arrived at it in a few weeks of simulated trial and error, on hardware that did not exist a decade ago. Two very different paths, the same physics, the same answer.

## THE COMPLEMENTARY PAIR

The conceptual frame that shapes current development is the recognition that large language models and robots are complementary in a precise sense. Language models know almost everything that humans have written, from quantum mechanics to the rules of etiquette. They lack a body. Robots have a body and actuators, but lack general semantic knowledge. The combination of a language model's knowledge base with a robot's physical presence creates a system more capable than either alone.

A robot programmed solely through traditional methods must be told what every object is and exactly how to handle it. A robot guided by a language model can receive an instruction like "put the plates away." Then, based on the model's understanding of kitchens and plates, infer a reasonable set of actions. The model does not guarantee that the robot will succeed. It expands the range of tasks the robot can attempt without custom programming.

This complementary relation is the engine behind most current commercial development. It is also why advances in language model capability translate directly into advances in robotic capability, even when the robot hardware remains unchanged. The pairing is the moment when machine intelligence stops being a disembodied reader of writing and starts to act in the same rooms that carbon-based life inhabits. The mind has read the cookbook. Now the body is being built that can stand at the stove.

## THE SAFETY & RELIABILITY QUESTION

Before any humanoid robot enters domestic or public environments at scale, it must meet standards that current systems do not reliably satisfy. The safety question is primarily about mundane but consequential failures that a 66-pound machine can cause—not the catastrophic scenarios of science fiction. Dropping a glass. Knocking a child over. Failing to stop when a person steps into its path. Falling over backward after an electrical fault.

1X's Neo did exactly this during one demonstration, falling backward after an electrical malfunction. A fall of that kind, in a room occupied by a small child or a frail elderly person, would be a serious incident. Companies acknowledge the concern directly. It is not dismissed as a distant problem. It is the engineering reality that precedes mass deployment.

Certification pathways for humanoid robots also do not yet exist in most jurisdictions. The regulatory frameworks that govern industrial robots, designed for fixed installations behind safety barriers, do not translate to mobile robots that share space with people. Automotive safety standards, developed for machines that travel on roads, address a different interaction profile. A new category of standards is needed. Standards bodies are in the early stages of developing them.

The absence of established standards is a regulatory inconvenience with real commercial consequences. It shapes what companies can sell and to whom. A hospital will not deploy an unproven robot in patient-care areas without certification. An insurance company will not underwrite a domestic robot without data on its failure rate and injury risk. The commercial pathway from early adopter homes to broad deployment runs directly through the development of these standards. That development takes years.

The capabilities grow each year. The gap between staged demos and reliable autonomous deployment is real and is well understood by everyone building these machines. The tools to close that gap are available. The question is not whether the gap closes. The question is how long it takes and what safety and regulatory infrastructure gets built along the way.

There is a longer view that frames the engineering one. For most of human history, the only bodies moving through the world were biological ones. Robotics is now attempting, in earnest, to build machine bodies that can operate in the same spaces. The process is slow, awkward, and expensive. Factory floors and test homes in Norway, California, and Shanghai are the sites of these early efforts. The bodies are clumsy in 2026, much as early aircraft were in 1910. The clumsiness observed now is likely an early chapter in the development of humanoid robotics.

## EMBODIMENT

Machine intelligence, in its most powerful current form, is a process that runs on silicon, consumes electricity, and produces text. It operates entirely within information space. It reads and writes, predicts and classifies, translates and generates. It does not mine ore, fabricate components, or move objects from one location to another... yet. It is powerful in ways no previous technology matches, and it is physically inert in a way that no hand tool is.

The robot changes this by serving as the interface between the information system and the physical world. When a large language model acquires a body, the outputs of its computations take on physical consequences. A planned action thus becomes a physical event. A prediction about the most efficient path through a warehouse becomes a set of steps on the warehouse floor. This coupling of machine intelligence with robotics marks a qualitative change in what machine systems can achieve in the material world.

*What does it mean for machine intelligence to acquire embodiment? Why does it matter beyond the obvious commercial applications? How does it set up the developments that follow?*

### Closing the Loop

Biologists use the term sensorimotor loop to describe the cycle in which an organism senses its environment, generates a response, and perceives the consequences of that response. The loop is not metaphorical. It is a physical circuit running from sensory organs through the nervous system to muscles and back, with a time constant of tens to hundreds of milliseconds. Life is organized around this loop. The nervous system did not evolve to contemplate, but to act.

Current machine systems have a partial version of this loop. They sense input in the form of text, images, audio, and other data structures, and they produce output. The output does not feed back into the environment in a way that the system directly perceives. A language model does not observe the consequences of its generated text. It does not feel whether its prediction was correct in a physical sense. Each conversation starts fresh.

A robot with a machine-intelligent brain is different. Model output, a planned movement or a chosen action becomes a physical event. Sensors report the outcome to the model. *Did the object move? Did the grip succeed? Did the step land?* This feedback serves multiple purposes. Within real-time systems, it corrects mistakes instantly. The loop closes.

The closure of this loop is what makes embodied machine intelligence different in kind from machine intelligence that processes information without physical consequence. The robot that reaches for an object and adjusts its grasp in response to slip detection is doing something a language model answering a question is not. It is acting in a physical world and adapting to that world's response in real time.

## Why the Body Matters

There is a theoretical argument about intelligence and embodiment that goes beyond robotics engineering. The argument holds that intelligence is not purely a computational property. It is partly a property of the interaction between a computational system and a physical environment. A brain without a body, or a system that processes information without physical consequence, develops a different kind of intelligence than one that has to navigate the continuous, unpredictable pressure of physical reality.

The evidence for this view comes partly from biology. The nervous system of the nematode worm *Caenorhabditis elegans*, with its 302 neurons, produces behaviors that are difficult to replicate in simulation because they are partly generated by the mechanics of the worm's body. The muscles, the cuticle and the fluid mechanics of the worm's motion through its medium are computational elements. The brain does not have to solve certain problems explicitly because the body's physics solves them implicitly.

In humans, the relationship between physical experience and cognitive development is clear during early childhood. Infants who manipulate objects develop object permanence and spatial reasoning more quickly than those with limited physical access to objects. The body is not simply the carrier of the brain. It is part of the cognitive system.

This view suggests that machine systems that acquire bodies may develop forms of understanding inaccessible to systems confined to text and image processing. A robot that handles physical objects under different conditions acquires an implicit model of how the physical world works—from accumulated experience of acting and sensing, rather than from explicit training labels.

Whether current machine architectures are capable of building this kind of implicit world model is an open question. Early evidence from robotic foundation models suggests that physical experience does improve performance on reasoning problems involving physical concepts. Models trained on robotic action data perform better on tasks that involve reasoning about physical consequences than models trained on text alone. The embodied turn may be producing a form of understanding that text-only training cannot.

## FROM INFORMATION LIFE TO PHYSICAL AGENT

Whatever else we call them, today's AI systems are doing something new. They process information at a scale no earlier technology could match. They generate responses. They can modify their own parameters. In sufficiently capable forms, they can assist in the design of their successors. Whether that adds up to something deeper is a question for another book. What matters here is that these capabilities are already reshaping the world they enter.

What they lack is physical autonomy. A machine system that cannot act in the physical world cannot mine materials, fabricate components, build the infrastructure it requires to run, or maintain the hardware it depends on. It exists at the pleasure of the humans who keep the data centers running, who supply the electricity and who manufacture the chips. Physical dependence is a constraint on any claim to genuine autonomy.

Robotics is the mechanism by which this constraint can be relaxed. A robot controlled by a machine system can, in principle, perform physical tasks, including maintenance and construction of robotic systems. A sufficiently capable robotic intelligence could, in theory, direct the mining of silicon, the fabrication of circuit components, and the assembly of computing hardware. This is speculative at the current state of development. It is also the direction in which the combination of machine capability and robotic dexterity points.

Current robots cannot do most of this, and the gap between current capability and self-replicating robotic systems is large. The robot is a product and something more. It is the mechanism by which machine intelligence begins to acquire the physical agency that, over time, would reduce its dependence on human-maintained infrastructure.

Biological evolution spent four billion years working out a body that could feed, repair, and replicate itself. Robotics has not solved any of those problems, but the humanoid robot is the place where the effort begins.

## SETTING UP WHAT FOLLOWS

This part of the book has followed robots from their current capabilities, through the near-term trajectory of deployment, to the deeper significance of machine intelligence acquiring physical presence in the world. Three things emerge from that sequence.

First, the gap between current demonstration and reliable deployment is real and well characterized. It is an engineering problem with a known structure and active solution paths.

Second, the combination of machine intelligence and robotic embodiment is the mechanism by which machine systems move from information life to a physical agent. The implications of that transition extend beyond any commercial application.

Third, the intelligence that emerges from physical interaction with the world is different in character from that which emerges from processing text. The embodied turn is the beginning of a qualitatively different relationship between machine systems and the material world they increasingly act on. It extends beyond mere extensions of current machine capabilities.

The chapters that follow examine what comes next when that relationship matures. They consider how people and machine systems inhabit a world they increasingly share, and what the terms of that cohabitation are.

✧✧✧

# CHAPTER 10
# CAN A MACHINE BE CREATIVE?

LONG BEFORE COMPUTERS painted their first recognizable image, humans wrestled with the notion of creativity: what it is, where it begins, and whether it resides within the individual or flows from something vaster that moves through us all. To ask whether a machine can be creative is, in the end, to peer into our own nature and wonder what creativity means to those who claim it.

To see how creativity manifests in a single human life, consider Ludwig van Beethoven. Beethoven's notebooks are a testament to obsession and creative struggle. Their surviving pages bristle with crossed-out bars and frantic marginalia. The same musical phrase is rewritten a dozen ways. It's as if the right sequence might emerge only after a hundred failed attempts. That now-iconic four-note motif that opens his Fifth Symphony was not a stroke of genius granted in a flash, but the result of years of grinding revision. The labor in those pages is a visible record of pain and persistence shaped by endless trial and error. The composer who would one day shake Europe with the most recognized notes in classical music lived for years inside uncertainty, trying, rejecting, and trying again until the storm in his mind finally cleared into music.

Yet the very same Beethoven, seated at a piano before an expectant audience, could conjure melodies that seemed to arrive from nowhere. His improvisations, witnesses swore, could move listeners to tears. Music poured out of him as if he were channeling some vast force. Those present described the experience as watching the weather form in real time. This duality—the painstaking craft and the effortless flow—reveals the complex landscape of creativity. The landscape resists a simple definition, whether the creator is made of carbon or code.

Creativity comes in at least two forms that can coexist in a single mind. The first is the grinding, iterative kind. This is where an idea is assembled piece by piece under conscious pressure. The second is called flow. It is the experience of effortless attention, a state in which skilled performance seems to run on its own. It is faster and more fluidly than deliberate reasoning allows. Athletes call it being "in the zone." Surgeons describe an entire operation passing in what feels like ten minutes. Writers know it as the morning that disappears between two cups of coffee.

For decades, we learned about flow through introspection. People described it, and experiments measured behavior before and after. Yet the inner machinery remained opaque. Psychologists identified several conditions that consistently preceded the state.

The task's challenge should closely match the performer's skill, so neither boredom nor anxiety dominates. If the task is too easy, the performer disengages. If it is too difficult, the performer freezes in self-conscious struggle. In a narrow corridor between those extremes, attention gathers into something that feels paradoxically effortless, though demanding.

Think of riding a bicycle at the speed at which the wobble disappears. Pedal too slowly, and you fall. Pedal too quickly, and your balance is overwhelmed by the effort. Only in a thin range in the middle does the bicycle seem to ride itself. The rider becomes a passenger on their own initiative.

The state enhances both productivity and meaning. People in flow report some of their most satisfying experiences. They lose track of time, of hunger, of the room around them. When the work ends, they often describe a strange post-flow tenderness, as though they return from somewhere.

A more recent line of inquiry used jazz improvisation to open that machinery to observation. Jazz suits this purpose because it produces a real-time, scorable record of creative decisions while the performer is still making them. The musician cannot hide. Every choice is audible, and the audience hears them in the order the musician makes them.

32 jazz guitarists, from near-beginners to highly experienced players, improvised solos over specified chord sequences while wearing specialized scalp caps that record electrical activity in the brain using a dense array of sensors. After each performance, the players rated the intensity of flow they experienced. Expert judges listened to the recordings and scored them for creativity. The team cleaned the brain signals of muscle noise and eye movement artifacts, then mapped them to their neural sources.

A pattern emerged from the higher-flow performances and challenged a popular assumption. Rather than showing intense, laser-focused brain activity that people might associate with peak performance, high-flow states correlated with reduced brain activity.

Flow is hyperfocus in reverse. It is a supervised relaxation of oversight, a state where the conscious supervisor steps back and lets a trained network run. Imagine a master sommelier tasting a wine. The novice analyzes, hunting for descriptors. The master notices the wine tastes of black cherry and then names it. The expertise sits closer to perception than to thought.

The most experienced players showed something the novices did not. During the flow, their auditory and visual cortices became active. Decades of practice built a specialized circuit for jazz guitar improvisation. Once they released the executive brake, this circuit could run on sensory processing.

Creative flow, on this account, requires two ingredients. Practice builds a dedicated internal network for the task at hand. Metacognitive judgment is needed to stop overthinking and let that network operate. The first takes years. The second requires a kind of self-trust that is itself a skill.

## THE BRAIN NETWORKS OF CREATIVE THOUGHT

If flow is one face of creativity, then divergent thinking is another. Flow describes a mode of production. Divergent thinking describes the capacity to generate many possible solutions to an open-ended problem, especially solutions that are distant from the obvious ones.

A standard probe for divergent thinking is the alternate uses task. A participant sees an ordinary object and lists as many unusual uses as possible. A sock can warm your feet. It can filter water, carry small tools, protect a delicate object, serve as a puppet for a tired toddler, or cover a doorknob to prevent a click as you slip out of a sleeping baby's room. The more original the list, the higher the creativity score.

People with stronger brain networks generate more original ideas. The networks do not belong to any single brain region. They span three distinct systems that ordinarily operate in opposition to one another.

The default network activates during spontaneous thinking. Mind-wandering, daydreaming and imagining scenarios not present in the environment. In the context of creativity, it is the act of generating. It's the unconstrained brainstorm phase, where possibilities multiply before evaluation begins. Think of it as the staff in the back kitchen, throwing ingredients on the counter and asking what might work.

The executive control network activates when attention needs to be focused and when thought processes need to be monitored. During creative work, it evaluates ideas. This network asks whether a proposed solution actually fits the problem and adjusts it when it does not. This is the head chef, walking the line, tasting each pan.

The salience network serves as a switch between the other two. It detects what deserves attention and routes processing accordingly. In creative work, it may toggle between generative and evaluative modes as a problem develops. This is the maître d' or head waiter, moving between the kitchen and the dining room.

These three networks normally counteract each other. When one activates, the others quiet down. People with higher creative ability appear to co-activate all three simultaneously, running generation and evaluation in parallel rather than in sequence. The creative brain is not differently organized so much as differently coordinated. It runs the kitchen, the line, and the floor simultaneously, and the result is a meal that arrives inventive and edible.

This pattern holds across populations. Professional jazz musicians improvising, poets drafting new lines, and visual artists sketching show convergent evidence of the same three-network engagement. Whether these connectivity patterns are fixed traits or trainable remains unclear. The neural substrate of creativity is plastic enough to change with expertise, as jazz guitarists with specialized sensory networks show. Whether practice can strengthen the links between networks is still to be established.

## CREATIVITY & ITS DISCONTENTS

Creativity has never belonged exclusively to recognized geniuses, and the public record has judged what counted as creative in its own time. Vincent van Gogh sold almost nothing during his lifetime. Many people regarded him as mentally unstable. Contemporaries dismissed his painting style, but it became foundational for the Expressionist movement that arose after his death in 1890. Samuel Beckett submitted his first novel to every publisher he could, and all rejected it. A publisher issued it in 1992, three years after Beckett died and twenty-three years after he received the Nobel Prize in Literature.

These cases are stories about rejection and something more. They illustrate a structural feature of creativity. Its products may arrive before the frameworks needed to evaluate them are available. Authorities executed Socrates for questioning the gods of Athens. He did not ask the wrong questions. His audience lacked the tools to recognize what he was doing. What looks inevitable in retrospect looks disruptive in the present.

Pablo Picasso's Cubism disturbed viewers accustomed to representational painting. Within a generation, it altered the vocabulary of music, literature, and architecture. The physicist Steven Weinberg, writing in 1972 about the Hubble constant, the rate at which the universe expands, found that two competing camps entrenched themselves in values of 50 and 100 kilometers per second per megaparsec, respectively. He proposed averaging them. Decades later, measurements showed the actual value is close to 70. Sometimes the creative act is simply refusing to choose between entrenched positions.

Arts and sciences share this structure without sharing their materials. Scientific creativity produces new descriptions of reality. Einstein's general relativity provided the mathematics that underlies GPS navigation. Artistic creativity produces new ways of experiencing reality. Cubism gave viewers a way of seeing objects simultaneously from multiple angles, making visible something that perspective-bound painting could only approximate.

When people asked Michelangelo how he produced his sculptures, he said the finished form existed in the marble. His task was to remove what was not part of it. Whether or not that describes what happens in a sculptor's mind, it captures something about the phenomenology of creative discovery. The answer was there, waiting. The work was to clear away what hid it.

## WHAT CREATIVITY IS NOT

Creativity is also often conflated with output. A painting is creative. The process that made it may or may not be. Defining creativity by its products, rather than by internal process, has a practical advantage. It allows the same standard to be applied to any system that generates novel, useful things, including machine ones.

The standard definition in psychological literature describes creativity as the ability to produce ideas that are, to some extent, *original and useful.* The definition says nothing about consciousness, intention, or substrate. It asks only whether the output satisfies the criteria. Agnosticism is not an oversight. It reflects a principled decision to measure what can be measured.

Applied to machine systems, the question becomes whether current systems produce outputs that satisfy the originality and usefulness criteria. The answer is, increasingly, yes, at a level competitive with average human performance.

The alternate uses task provides one benchmark. In a systematic comparison, large language models completed the same task as human participants. The machine systems performed better than the human average by semantic distance metrics, which measure how conceptually far a proposed use is from the object's conventional function.

On subjective ratings, where trained judges scored responses for cleverness and unexpectedness, the machine systems again exceeded average human performance. In roughly 7% of human trials, participants produced responses so common or incoherent that judges assigned the lowest possible scores. The judges gave the machine systems no scores in that range. The process swept the bottom of the distribution clean.

The human-machine difference appears to rest on executive function. The weakness in human creative performance concentrates at the bottom of the distribution. Ordinary people generate ordinary ideas because their inhibitory control fails to suppress the most obvious associations. The first answer that comes to mind when asked about the use of a sock is often to *put it on your foo*t. Machine systems do not struggle with that kind of failure. They do not have motivational lapses, fatigue, or anxiety about being judged. They execute the task computationally.

At the top of the distribution, the picture reverses. The most creative individuals still match or exceed machines on semantic distance scores. The human genius is not yet outperformed. What the machine does is compress the gap between the ordinary performer and the extraordinary one, making the mean rather than the maximum competitive. The ceiling has not moved. The floor has risen.

## HOW DIFFUSION MODELS GENERATE THE NEW

While the alternate uses task illuminates behavioral creativity, a parallel line of inquiry asks a more mechanical question. *What structural properties of machine systems give rise to outputs that are genuinely novel rather than memorized?*

Diffusion models, the architecture behind image-generating systems like DALL·E and Stable Diffusion, were designed to reproduce training images, not invent new ones. They work by adding noise to an image until it dissolves into incoherent pixels, then learning to run that process in reverse. Given a field of noise, they reconstruct a plausible image. Imagine milk being stirred into coffee. Then imagine it runs backward until the milk gathers itself into a tidy white blob at the surface.

In practice, however, the models blend elements across training examples, producing coherent images that do not match any single source. A prompt for "a cat wearing a top hat in the style of Vermeer" returns an image that exists nowhere in the training data, in a style the painter did not live to see. This capacity for recombination puzzled the field for years. *If the models were just reversing the noise-addition process, why did they produce anything new at all?*

Physicist Giulio Biroli called this the paradox of diffusion models. If they worked perfectly, they would memorize. Instead, they create. The answer, presented at the International Conference on Machine Learning in 2025, lies in two computational constraints that were long dismissed as mere technical limitations. Mason Kamb, a graduate student in applied physics at Stanford, and his supervisor, Surya Ganguli, identified these constraints as locality and translational equivariance.

Locality means that during reconstruction, the model attends to a small patch of pixels at a time rather than the entire image. Translational equivariance means that if an input is shifted slightly, the corresponding output shifts. This preserves coherent structure. Since each patch is generated without knowledge of its final position in the final image, the model cannot simply replay a stored template. It must assemble the image from locally consistent decisions, with each patch fitting its neighbors, without any global blueprint anywhere in the process.

Kamb recognized the pattern. He had been studying morphogenesis, the process by which embryos self-assemble. Patterns show how groups of cells can organize into distinct organs and limbs through purely local signaling, without a central coordinator or a finished plan. Individual cells respond only to their immediate neighbors. The system produces a functional body. Occasionally, it goes wrong and produces hands with extra fingers.

The first machine-generated images to flood the internet shared the same signature flaw. Figures with too many fingers, faces with too many teeth, hands that branched into one extra knuckle. This is the failure mode of a bottom-up, locally constrained system. The hypothesis was that locality and equivariance were the generative mechanisms behind machine creativity—not incidental properties. The same architecture that makes the model fail to count fingers correctly is also what lets it produce anything new at all.

To test the hypothesis, Kamb and Ganguli built a system they called the equivariant local score machine. It is not a trained model. It is a set of equations that predicts the output of diffusion models based solely on the mathematics of locality and equivariance. They ran the same noise-corrupted images through the equations and several standard-trained models.

The equations matched the trained models with an average accuracy of 90%. Ganguli described this figure as unheard of in machine learning. Creativity, in diffusion models, does not emerge. It occurs as a deterministic consequence of architecture. The model is created because it cannot see the whole picture at once, just as a stained-glass artist who works one pane at a time produces an image that no single drawing on the studio floor predicts.

## PLATO'S GEOMETRY & THE CHATBOT

A different kind of creativity probe comes from mathematics education. Plato's dialogue *Meno* contains what may be the earliest documented experiment in mathematics instruction. Socrates guides an uneducated boy, step by step, to discover how to double the area of a square. The boy initially doubles the side length, which produces a square four times the area, and then is led to see that the correct solution is a square whose side equals the original square's diagonal. The dialogue raised a question that has occupied philosophy for centuries. *Does the boy retrieve knowledge he already holds, or does he construct new understanding through dialogue?*

A version of this problem was posed to ChatGPT-4. The model was expected to recognize the scenario and reproduce Socrates' geometric solution, since the *Meno* dialogue is among the most widely reproduced texts in Western philosophy and would appear in virtually any large training corpus or dataset.

The model did not reproduce the known solution. When presented with the problem of doubling a square, ChatGPT chose an algebraic approach, working with equations rather than a geometric construction. It resisted prompting toward the geometric method. It insisted on algebra even when the team expressed dissatisfaction with approximate answers. Only when the model was told they were disappointed that it could not provide an "elegant and exact" answer did it produce the geometrical construction.

They also posed a variant. Double the area of a rectangle while preserving its proportions. Despite now knowing their preference for geometric solutions, ChatGPT reverted to algebra. When pressed further, it claimed that a geometric solution was unavailable because the diagonal of a rectangle cannot be used to double its size. The claim about the diagonal is mathematically correct. The claim that no geometric solution exists is false. A different geometric construction works.

The model was not retrieving a stored false answer. It was improvising on the basis of its previous responses, the way a student might generalize incorrectly from a partial correct insight. The behavior is learner-like. It's a blend of retrieval and on-the-fly reasoning rather than pure recall. The system had walked itself into an error of the kind that a clever undergraduate makes at the chalkboard.

These results do not establish that ChatGPT solves mathematical problems the way humans do. Large language models predict word sequences. They do not manipulate symbolic representations in the way a proof-checking system would. From the perspective of users interacting with the model, what appeared at the surface was improvised problem-solving with human-recognizable errors. The model showed something that, functionally, resembled working at the edge of its competence, pushing into territory it understood only partially.

This is similar to the educational concept of a zone of proximal development. The zone is the gap between what a learner already knows and what they might reach with guidance. Students working with a machine on mathematical problems should treat model-generated proofs as objects to verify, not as proofs to accept. Understanding and evaluating machine reasoning is becoming a core mathematical skill, one that mirrors the critical habits mathematics education has always aimed to cultivate.

## ASSEMBLING THE NEW FROM PIECES OF THE KNOWN

Human and machine creativity also share a structural feature that neither side of the debate about machine creativity has fully confronted. Recombination. On this point, human minds and machine models look more alike than either camp has been willing to admit.

Beethoven's notebooks show that his great themes were assembled from fragments, tested in dozens of combinations before one cohered. The jazz guitarist in the flow state draws on years of stored patterns, reconfiguring them in the moment. The diffusion model patches pixels from its training distribution into novel configurations through local compatibility. ChatGPT draws on a training corpus to generate text that no single source contains.

The critical distinction may not be the act of recombination, which appears common to human and machine alike. It may be the mechanisms that guide selection among possible recombinations. In the human creative brain, the default network proposes, the executive network evaluates, and the salience network switches between them. The quality of the output depends on the richness of the generative material, the rigor of the evaluative filter, and the speed at which the system can cycle between the two. These are properties that can be trained, as Beethoven's notebooks and the jazz guitarist's years of practice demonstrate.

In machine systems, analogous functions are performed by learned weights, attention mechanisms, and sampling parameters. The generative material is the training corpus. The evaluative function is built into loss minimization during training and into alignment procedures afterward. The cycling between generation and evaluation happens at the level of tokens, not hours or years of deliberate practice. The artist who refines a piece over six months and the model that generates a draft in six seconds are doing something with the same shape, conducted at radically different time scales and through radically different machinery.

The comparison does not suggest that the machine is creative in the same sense as Beethoven. Creativity, viewed functionally, is a class of information-processing behaviors rather than a uniquely biological process. Generate candidates, evaluate them against criteria and retain those that satisfy the novelty and coherence requirements. Any system that performs those steps with sufficient richness of generative material and sufficient rigor of evaluation will produce outputs that meet the standard definition of creative. Whether we call that "true" creativity depends on what we mean by the word. That is a philosophical question rather than an empirical one.

What the empirical record shows is this. Machine systems produce outputs that routinely exceed average human performance on divergent thinking tests. Their architecture generates genuine novelty through local recombination rather than memorization. They improvise in ways that produce recognizable human-like errors when working at the edge of their knowledge. The creative brain and the generative model operate differently at the mechanistic level. At the level of behavior, the distance between them narrows.

## THE LIMITS OF THE COMPARISON

None of this settles the deeper question. Behavioral similarity does not imply identity of process, and identity of process is not required for creative outputs. A camera and a human eye form images. They do so through entirely different mechanisms. What matters for any particular purpose is which comparison is relevant to that purpose.

For artists, the relevant comparison may be experiential. The difference between a work made through years of embodied practice and one generated by a statistical model trained on the artifacts of that practice. The robot sees only the last layer of a painting. It cannot see the sketch that was abandoned, the morning the artist could not work, the year they spent in a hospital that taught them how to paint loss. The surface of a creative work can be reproduced. The history of decisions that produced it cannot.

For users of creative outputs, the relevant comparison may be more functional. *Does the output achieve what is needed, and does it matter how it was made?*

For many practical purposes, it does not. A presentation deck, a stock illustration, a placeholder melody for an animated short.

For the purposes that touch on meaning, expression, and communication between persons, it probably does. A condolence note written by a machine and a condolence note written by a friend say the same words. Only one of them carries the friendship.

For cognitive scientists, the relevant comparison is mechanistic. *Does the machine system use the same processes as the human brain, and if not, does the difference matter for explaining the outputs?*

The diffusion model suggests that some outputs can be explained mechanistically, with analogs in biological self-assembly. Whether those analogs reflect a genuine shared logic or only a superficial resemblance requires more investigation than either neuroscience or machine learning can provide at present.

The question of whether the machine is creative, posed plainly, turns out to contain several different questions.

*Is the machine capable of generating novel, useful outputs?* Yes.

*Does it do so through processes that resemble human creativity at the level of neural mechanisms?* Partly, and in ways that are still being mapped.

*Does it experience anything while doing so?* Unknown, and the question may be unanswerable with current tools.

*Does the answer to any of these questions determine how we should treat machine-generated creative work?* That is a question for culture, law, and philosophy, not neuroscience.

The systems this book examines have begun to make things. Some of those things are competent. Some are striking. A few, in the right hands, are beautiful. The hands that hold them, for now, are still human. What we choose to make with these tools, and what we choose to make of tools that now make alongside us, is the work of the years immediately ahead.

✧✧✧

# CHAPTER 11
# THE CHANNEL BETWEEN MIND & MACHINE

BRAIN-MACHINE INTERFACES TURN hope into reality for people with neurological damage, allowing the mind to reach past broken pathways. A person paralyzed by a spinal cord injury moves a prosthetic limb or a computer cursor—using only the silent language of their motor cortex. Another, silenced by brainstem damage, regains speech as the sensorimotor cortex links directly to a speech synthesizer. Where tissue falls short, engineers craft a detour, building new routes.

Medicine applies this principle across many devices. Eyeglasses bend light to correct vision. A cochlear implant converts sound into electrical pulses, allowing the auditory nerve to receive signals when cells fail to function. A cardiac pacemaker restores heart rhythm by delivering timed electrical impulses. Each device replaces damaged tissue with engineered signals. A brain-machine interface uses the same logic, substituting for the cortex or its connections to the muscle. Each step along this ladder, from lens to cochlea to heart to cortex, brings engineered matter deeper into evolved tissue.

Engineers restore function through silicon rather than axons, but the aim is the same. The goal is to return a person to ordinary engagement with the world. As brain-machine interfaces advance, the line between repairing and augmenting the brain shifts. The distinction between therapeutic restoration and elective enhancement becomes a matter of degree.

A person sits at a computer and moves the cursor with only their thoughts. Electrodes on the scalp record faint electrical signals from the cortex. These signals reflect many overlapping processes: the visual cortex tracks the screen, the auditory cortex registers background noise, the prefrontal cortex weighs decisions, and the motor cortex prepares possible movements. Translating this complexity into cursor movement requires distinguishing one pattern from many. Electrodes alone cannot achieve this. Machine intelligence learns the specific patterns in an individual's brain signals and converts them into machine commands at a speed that feels immediate.

Without this intelligence, electrodes record only noise. The signals have no clear content. With the intelligence, thoughts become actions, and the line between intention and device shifts. For the first time, a boundary once at the skin moves inside the skull. This closeness changes what it means to connect mind and machine. It also changes what it means to have a body. The technology, which for most of its short history has run on its own substrate, now extends into ours.

## WHAT THE BRAIN DOES WHEN IT SPEAKS

Speaking feels effortless. At the neural level, nothing could be more complex. Even a simple sentence like "please pass the salt" recruits a network that stretches from the front of the brain to the back of the throat. This network coordinates with millisecond precision.

The simple act of speaking conceals an intricate choreography inside your head. Long before a word leaves your lips, the brain is already at work. Deep in the prefrontal cortex behind your forehead, decisions spark. Ideas take shape; intentions form. The plan travels to Broca's area, where raw thoughts are shaped into coherent sentences. Wernicke's area, ever vigilant, reviews the draft, ensuring the words make sense. Only then does the baton pass to the sensorimotor cortex, which conducts a symphony of over a hundred muscles in your chest, throat, and mouth. Each muscle moves in perfect synchrony, transforming silent intention into living sound.

As the speaker makes sound, the auditory cortex monitors the voice and sends feedback to Wernicke's area. This feedback lets the brain correct speech in real time. The loop keeps speech fluent even when the content changes. It also explains the small delay you might notice when you say a word that does not quite land. The system works in two directions. Output and monitoring run in parallel, like a skilled musician playing a note while adjusting the tuning.

Damage to the sensorimotor cortex prevents motor commands for speech from reaching muscles. The intention remains, but execution fails. The words persist within the mind. Yet, the ability to express them is lost.

Ann was 30 when a sudden stroke rewrote the rules of her life. Overnight, she lost the ability to walk and to move the muscles that shape words. Yet inside, her intellect and language remained untouched. Her thoughts still raced. Her inner monologue was as vivid and articulate as ever, but the path from mind to mouth had been severed. For 18 years, Ann's voice lived in silence. Her only means of communication a painstaking process—typing out each letter, one at a time, using a head-tracked keyboard. At her fastest, she managed 14 words per minute, a mere fraction of the speed at which she once spoke, and a world away from the rhythm of ordinary conversation.

## DECODING ANN'S VOICE

A team at the University of California, San Francisco, working with Berkeley engineers, restored that speed. They placed an array of 253 electrodes directly onto Ann's sensorimotor cortex. This region normally sends movement signals to the speech muscles. The method placed sensors on the brain's surface without penetrating the brain. It is less invasive than microelectrode arrays placed within the brain.

Ann formed sentences silently using a 1,024-word vocabulary and mouthed the words. The electrodes recorded her brain activity. They mapped when and where her cortex sent speech commands, even when her muscles failed to respond. These recordings went into a speech-decoding computer program trained to pair Ann's brain signals with the correct words.

Personal details shaped the process. Pre-stroke recordings of Ann's natural voice, taken from a wedding video, were uploaded so the output would sound more like her actual voice than a generic synthetic one. When the system finally spoke, it used the voice of the woman she had been at 29.

After training, Ann can send speech at near-conversational speed. Her brain signals pass through the decoder, producing sound at 78 words per minute. This is over five times faster than her typing rate. The system stays accurate even as her thoughts get faster. Earlier devices had robotic, stuttered speech. This system gives real speech rhythm, including rises and falls for questions and jokes.

The result is communicative and personal. Ann hears her own voice again as an immediate reality produced by her neural activity. The technology does not restore the biological pathway. It constructs a new one, routed through silicon rather than muscle.

The process resembles a river diverted by a rockslide. Conventional medicine removes the obstruction. A brain-machine interface creates a new channel, allowing the flow to resume along a different path. The water is unchanged. The riverbed is new. The person supplies the water, the intention, the meaning. The machine supplies the riverbed.

## THE PATH TOWARD WIRELESS DECODING

Ann's case used implanted electrodes connected by wires through a small port within her skull. This approach imposed engineering constraints. Wires passing through bone increased the risk of infection, restricted movement, and limited long-term use outside supervised settings. Engineers are now developing wireless transmission, in which small implanted devices encode neural signals and transmit them via radio frequency to an external receiver.

At New York University, wireless speech decoding was demonstrated using the same method employed in Ann's case, with electrode arrays placed on the brain's surface rather than inserted into tissue. This approach simultaneously records from large populations of neurons. It captures the patterns involved in speech production. The system incorporates deep learning structures. These include recurrent networks and transformer-based models. It used these to decode speech from neural recordings across a range of subjects.

Speech decoding accuracy does not depend strongly on whether signals are recorded from the left or right hemisphere, even though the left has traditionally been considered the speech-dominant side. People who have sustained damage to the left hemisphere, as often occurs in stroke-related aphasia, may still provide usable signals from the right hemisphere for a speech decoder.

The density of electrode arrays also matters less than expected. Lower-density arrays perform comparably to higher-density arrays in speech decoding accuracy. Using lower-density arrays means simpler devices, lower cost, and potentially greater accessibility. A technology that requires 300 electrodes per square centimeter will reach only a few dozen patients at a few elite centers. A technology that uses 30 electrodes per square centimeter can reach many more.

The team made the neural decoding pipeline from this work publicly available. This allows other groups to build on the architecture rather than reconstruct it from scratch. Open infrastructure speeds up progress, as open-source software did in computing. Each new team that contributes refinements and data improves the shared baseline, so what is today a lab tool moves closer to a clinical one.

The trajectory leads toward systems small enough to fit entirely within the skull, wirelessly transmitting. Subjects would carry no external hardware. The interface would operate continuously, going beyond explicit sessions. Continuous operation raises concerns about privacy and autonomy. No previous medical device has set a precedent for this. A device that monitors neural activity around the clock can access information about cognitive states, intentions, and experiences that the user has not chosen to share and may not realize they are producing.

## HOW MACHINE INTELLIGENCE LEARNS BRAIN SIGNALS

Brain-machine interfaces have another core challenge: neural signal variability. The brain has around 86 billion neurons that send electrical impulses. When a person thinks, moves, or rests, neurons create patterns in certain regions. The motor cortex makes unique patterns for each movement intention. The visual cortex changes patterns based on what someone sees. No two brains make the same patterns for the same task. Even in one brain, patterns shift as the person learns, tires, feels, or ages.

Engineers are not translating fixed languages. Instead, they work with a language that changes daily. The word for "left" on Tuesday may differ from Wednesday. The system has to learn one person's dialect for a day, then adapt as it shifts.

Machine learning tackles this challenge. Engineers train a machine intelligence system on an individual's brain activity, building a model that links patterns of brain activity to intended actions. During early training, a person imagines moving a hand left or right while the system records brain activity. The system analyzes electrical patterns. It identifies features that distinguish left from right and begins to predict intended movements from new recordings. Early predictions are noisy. With more examples, the signal becomes clearer.

Training continues beyond the basics. The brain changes as a person learns, tires, or experiences stress. For a brain-machine interface to remain effective, its artificial intelligence must also adapt. Reinforcement learning provides this adaptation. The system receives feedback about its performance and updates how it interprets signals. When the machine errs and the user corrects it, the system incorporates that correction. Over time, the user's intentions and the device's actions become more closely aligned.

Deep learning, which uses neural networks with many computational layers, advances this further. These networks identify features in brain signal data at multiple levels of abstraction, just as the human visual system identifies edges, then shapes, then objects, then scenes. Lower layers detect simple regularities in the electrical record. Higher layers combine those regularities into representations of intended speech sounds, movements, and action categories. The power to decode brain signals associated with intended speech, as seen in Ann's case, relies on deep learning networks capable of resolving fine distinctions in the temporal structure of sensorimotor cortex activity at a few-millisecond resolution.

## NEURAL DECODING FOR MOVEMENT

For people with paralysis, the capacity to control an external device through thought addresses fundamental losses of independence. Neural decoding models, machine intelligence systems that predict intended movement from motor cortex signals, make this possible.

The motor cortex is the brain region controlling movement. It sends out detailed instructions for every motion: where to move, how fast, how hard, and in which direction. Computer models can be trained to read these instructions from brain recordings and figure out exactly what movement someone is trying to make. In studies with people who couldn't walk, computers read their brain signals and used them to control robotic machines worn around the legs. Those who hadn't walked in years could take steps again.

The granularity of control available through these systems goes beyond simple movements. A prosthetic arm controlled via a neural decoder can adjust grip pressure, vary movement speed, and respond to changes in intended action in real time. A patient at a Pittsburgh trial used such an arm to pick up a chocolate bar, unwrap it, and place it in her mouth. It was the first time in nearly a decade she had fed herself. The arm did not operate through switches or buttons. It responded to the same signals that would, in an intact nervous system, direct the biological limb.

With experience, users' brains adapt to the interface. The plasticity that improves control of biological limbs also forms the signals sent to a prosthetic. Early control is imprecise, but practice organizes neural signals and strengthens their association with specific intentions. Adaptation occurs on each side of the interface. The idea originally was to create a tool for the patient to master. Instead, the tool also adapted and improved with the patient. The two sides learn together. Progress depends on the combined efforts of humans and machines.

This merging is part of the brain-machine interface work that surprises people new to the field. The image of a static device receiving commands from a brain is wrong. What emerges instead is a partnership in which the algorithm and the cortex co-adapt over weeks and months, neither holding still, each pulling the other into closer alignment. The brain and the machine learn each other's languages, slowly, the way two species sharing a habitat eventually arrive at signals each can read.

## COGNITIVE ENHANCEMENT THROUGH THE INTERFACE

Brain-machine interfaces aren't only for repairing lost abilities. They can also boost or enhance the abilities you already have. Some systems do more than read brain signals. They also send small electrical pulses into the brain to change activity in targeted areas.

Experiments target the hippocampus and surrounding structures. Delivering stimulation when encoding of new information is weakest improves the accuracy of later recall. The system monitors neural activity in real time. It can process when the brain is not fully focused on taking in new information. When it spots these moments, it sends a small, carefully timed electrical pulse to the brain area responsible for memory. This pulse helps the brain become more ready to store and keep the new information.

In other words, the device gives the brain a gentle boost at just the right time. It makes it easier to remember what is being learned. It's like the difference between writing a note on damp paper and on dry paper. The hand is identical, but the trace it leaves changes.

Attention is another area that can be improved with these systems. Machine intelligence can monitor changes in brain activity to detect when someone's mind is wandering or they're feeling tired. When it spots these patterns, the system can give a soft nudge to help the person refocus. This kind of live feedback could help people who have trouble with attention, such as those with ADHD.

These applications share a common logic. Machine intelligence reads the brain's internal state, identifies a condition that can be improved, and intervenes to shift the state in a targeted direction. The intervention delivers specific modulation to a defined circuit at a precise moment, guided by the ongoing neural signal. The patient's brain, not the clinician, determines the timing of the intervention.

## BRAIN-TO-BRAIN COMMUNICATION

The same technology that transmits brain signals to computers can also transmit them to another person's brain. In some experiments, one person's brain activity is decoded and then delivered to another person's brain using noninvasive scalp stimulation. The recipient can respond to information not presented through ordinary senses, as the signal travels directly from one brain to another through a digital relay.

The demonstrations remain limited in scope. Experiments use simple signals, controlled lab conditions, and coarse transmission.

Decoding, digitizing, and transmitting neural activity make it possible to influence neural activity in another brain. The quality of these brain-to-brain signals improves as the computer programs that read and send the signals become more advanced. In simple terms, clearer technology and smarter software make the information sent between brains more accurate and reliable.

The implications run in several directions. Brain-to-brain communication might become partially possible across language barriers if the transmission conveys pre-linguistic intention rather than language itself. This means the conceptual core before the speaker selects a word. People who cannot speak or write can transmit information directly, without any expressive channel.

Cooperative problem-solving might combine two people's neural states in ways that exceed what either can articulate verbally. Two surgeons working on the same case, two musicians performing together, and two pilots flying in formation already share a state through tone, glance, and gesture. A neural channel deepens what already exists. In this scenario, the machine becomes the medium through which two minds meet directly, without sound or print.

Each of these possibilities carries risks that we should not consider incidental. The ability to read and transmit neural signals also enables surveillance of intention. Protecting privacy at the level of thought requires safeguards that no legal or technical framework currently provides. Society has not yet confronted the challenge of protecting thoughts and intentions that remain unspoken. As technology advances to the point of potentially accessing and transmitting neural activity, the need to address the rights and privacy of our innermost thoughts becomes increasingly urgent.

## WHAT HAPPENS COGNITIVELY WHEN HUMANS TALK TO MACHINES

Machine intelligence systems also change the way people converse, cognitively and socially. These differences affect how the brain processes information during and after the interaction.

A machine system lacks a face. It does not produce the nonverbal signals that human interaction continuously generates. It also misses the spontaneous, unpredictable qualities of another mind. When people interact with machine intelligence, their brains respond differently than when they talk with other humans. Instead of relying on fast, intuitive social skills, the brain relies more on careful, logical thinking. This changes how we process and understand these conversations.

Deliberate processing suits many cognitive tasks. People seeking information about a medical condition from a machine system benefit from the absence of social dynamics, hierarchy, and emotional pressure that complicate the same conversation with a physician. The machine does not convey impatience or judgment. It does not carry the implicit signals of a professional relationship in which the patient feels the visit's running clock. Many patients report asking their machine system questions they would not ask their doctor, due to the fear of sounding foolish. The machine's lack of social presence, in such cases, enables the conversation.

Reliance on a system that responds quickly, accurately, and without friction reduces the demand on the user's own cognitive systems. Habits of mind that require practice, such as constructing an argument, evaluating competing evidence, and sitting with ambiguity, are exercised less often when a system handles them on demand. The cognitive trace of these habits weakens when the conditions for practice diminish.

Over time, the brain adapts to conversing with machine systems. Neural plasticity does not stop at the keyboard. The interaction itself becomes an environmental condition. The brain responds to it as it does to any sustained environmental condition, reconfiguring in response to what the environment demands. The first life slowly takes shape in its conversations with the second.

## THE BROADER MERGING

Right now, brain-machine interfaces are mainly used as specialized tools for specific medical needs. In the future, these devices are likely to become part of our everyday environment. Engineers are working on technology that can pick up brain signals and also send gentle electrical signals back to the brain, creating a loop. This means the device can continuously monitor the brain's activity, recognize when something could be improved, and give the brain a boost at just the right moment. This kind of back-and-forth communication with the nervous system is something medicine hasn't been able to do before. It could change how we treat many conditions.

Current devices deliver stimulation at fixed parameters. They are set once and rarely adjusted. Next-generation devices will adjust stimulation in real time based on the neural signals they continuously monitor. The system responds to the brain. The brain responds to the system. This creates a loop that operates faster than any conscious adjustment by either the clinician or the patient.

At the cognitive level, the merging of mind and machine already begins through the tools people use constantly. Language models draft and revise, search systems retrieve and rank, and translation systems change across languages in real time. These tools do not read the brain directly. They interact with thought through the interface of language and action. This effect on cognition is indirect but real. New neural interfaces now make the same interaction explicit. The machine reads the brain, responds to it, and potentially modifies it.

Whether mind and machine will merge in a deeper sense remains an open question. In functional terms, some early forms of merging are already underway. The question concerns the type of merger. One possibility is that humans retain direction, and machines serve as increasingly capable prosthetics for intention. Another is that the distinction between what the human intends and what the machine produces becomes difficult to locate. The answer depends on design choices, regulatory systems, and a continuing account of what human cognitive autonomy means and why it matters. The technology this book describes did not arrive to replace us. It arrived to entangle with us.

The asymmetry between reading and writing is significant as well. Devices that only read neural signals and translate intentions into outputs extend human agency. The user remains the source of intention. The device is a conduit.

Devices that read and write, receiving neural signals and delivering stimulation in response, create a feedback loop that includes an external processing system. Whether the system amplifies or shapes the user's intentions depends on its design and objectives.

Prosthetics that restore lost function serve clear medical purposes. Users profit directly, and the device's objective corresponds to the user's welfare as the user defines it. Enhancement devices designed for elective use in healthy individuals introduce more complicated incentives. Commercial entities that develop and sell these devices pursue interests in engagement, data collection, and product differentiation. The same interests have shaped the design of social media platforms over the past two decades.

Regulatory systems for implantable medical devices originated for therapeutic applications. Their adequacy for enhancement applications, or for the hybrid case of a therapeutic device whose long-term effects include cognitive changes, remains untested. The medical device framework asks if a device is safe and effective. The harder question is whether a device that is safe and effective for one purpose is also safe and effective for the user's whole life. The questions are not new. The instruments that make them urgent are.

People with paralysis, aphasia, and severe movement disorders already benefit from these technologies. Ann speaking in her own voice is the present reality, not a thought experiment. Regulatory and ethical analysis must begin while design choices remain open. Once deployment patterns are established, changing them becomes difficult. The boundary between brain and machine is shifting. The task now is to determine where, and on what terms, a new boundary settles. Brain and machine are no longer running in parallel. They share a circuit. What the next decades may describe is a negotiation, not a contest between them—conducted in code and tissue at once, over how the two will live inside the same body.

✧✧✧

# CHAPTER 12
# LIVING LONGER WITH A PARTNER THAT DOESN'T AGE

THE DREAM of living longer is as old as humanity itself. In the flickering candlelight of Egyptian tombs, priests carved prayers onto stone, asking for endless days. Greek myths spun stories of mortals granted unnatural ages—some blessed, others cursed—reminding listeners that immortality could be both gift and burden. Medieval alchemists chased the philosopher's stone, while Spanish explorers sailed into the unknown, lured by rumors of a fountain that could wash away the years. Across continents and centuries, the yearning to push back death has braided itself into every strand of human culture, as persistent as the ticking of a clock.

What is new is the speaker. Today, technologists address human longevity. They frame their predictions in terms of computers, models, and parameters. Two recent statements give the flavor of the moment.

Dario Amodei, the chief executive of Anthropic, wrote in a 2024 blog post that machine intelligence might double the human lifespan to 150 years. He proposed a compressed timeline to match the pace of technological development and pointed to the doubling of average life expectancy across the 20th century as a precedent. At the World

Economic Forum in Davos, he stated that the doubling might be achieved within five years.

Ray Kurzweil, whose predictions about computational progress have a mixed record, claimed that machine systems could pause aging by 2032. He suggested this might be possible through medical nanobots capable of repairing cellular damage and delivering drugs to specific tissues. The same author also proposed digital backup of the human mind as a route to functional immortality.

These claims circulate widely. They are taken seriously, in some quarters, simply because they are made by people of consequence in the technology industry. The best approach is to clarify what is actually possible before accepting or rejecting these claims outright. The central questions are: *What can machine systems truly accomplish in aging research right now? What advances might be possible as the field evolves? What remains firmly outside any evidence-based forecast?*

Failing to distinguish these questions leads to much of the current confusion.

## TWO DIFFERENT GAMES

In 1824, the average life expectancy in the United States was about 40 years. Today, it is approximately 76-78. This gain of nearly four decades is among the largest improvements in human outcomes.

It resulted from three sources: reductions in infant and child mortality, control of infectious diseases, and improved management of injuries. These gains did not come from slowing the aging process itself. It came from removing the causes of early death. Surviving childhood, infection, and injury, people still age on the same schedule as before.

Picture life as a winding trail etched across a wild landscape. For most of human history, the path was treacherous and shadowed by cliffs, swift rivers, lurking predators, and invisible infections. Many travelers never reached the far end. The 20th century, with its advances in medicine and public health, cleared away most of these dangers. Now, the majority make it to the steep, final ascent. The real triumph is getting there at all. However, slowing the climb up that last incline—that is a different challenge altogether.

The challenge now is the gap between lifespan or how long people live and healthspan or how many years are healthy. A Mayo Clinic study finds the U.S. healthspan lags lifespan by 12.4 years. Americans spend more of their added years managing chronic diseases such as cardiovascular disease, type two diabetes, cancer, osteoarthritis and dementia. 20th-century longevity extended illness as much as health. The rise in life expectancy is a triumph but highlights the quality of the added years as a pressing concern.

S. Jay Olshansky, an epidemiologist at the University of Illinois Chicago, published a careful analysis in *Nature Aging* on the implausibility of a near-term longevity revolution. He puts the distinction plainly. The game played over the past century aimed to eliminate premature death. The game now involves the biological process of aging itself, which is not yet understood well enough to be reliably modified.

Today, one in six Americans is 65 or older. By 2034, this group will outnumber those under 18 for the first time. This demographic change brings more chronic disease, overwhelming a health system unprepared for such a scale.

## WHAT MACHINE SYSTEMS ACTUALLY DO IN AGING RESEARCH

The places where AI adds meaningfully to aging-related science deserve to be described accurately rather than with fanfare. Large language models and other machine-learning systems accelerate three kinds of work, all of which the earlier chapters on research, drug discovery, and science covered in detail.

The first is reading through the published literature. The quantity of this literature is far larger than any individual scientist can hope to keep up with. The second is generating candidate hypotheses from biological datasets that no single lab could analyze on its own. The third is screening libraries of chemical compounds for those that might have a useful biological effect.

A useful analogy is the difference between a working chef and a kitchen appliance. The appliance can chop a hundred onions in the time it takes the chef to chop two. The appliance is genuinely useful. It changes the economics of preparation. It does not change what a meal is. It does not invent new cuisines. The chef still has to decide what to cook. The cooking still has to be done. Machine systems in aging research are powerful appliances in the kitchen of biology. They are not a substitute for the cook.

Aging happens because of many changes in our bodies over time. These changes fit into roughly 9 main types. These are: damage to our DNA, telomere shortening, changes in how our genes are switched on or off, problems with how proteins are handled in cells, issues with how cells sense and use nutrients, trouble with the cell's power plants, older cells that stop working but don't die, problems with how cells communicate, and running out of new stem cells. Each of these is a significant area of research. All of them are connected in complicated ways.

Machine learning and artificial intelligence can help by sorting through huge amounts of data, spotting patterns, and suggesting which treatments might be worth testing. These tools speed up research. They do not offer a simple fix for aging.

There already exist compounds that extend maximum lifespan in rodents by 25% to 50% with limited adverse effects. Rapamycin, which alters how cells sense nutrients, is among the most studied. The lifespan extension is reproducible across multiple independent labs.

The result does not translate straightforwardly into human terms. Rodents and humans age through overlapping but not identical mechanisms. Aging also happens on vastly different timescales, with different baseline immune and metabolic profiles. A drug that extends a mouse's maximum lifespan by 40% is a significant result in mammalian biology. There is no evidence that the same effect will appear in a human being.

A useful way to picture the gap between mouse and human is to think of a model railway. A child can build a small replica of a city's subway system in a basement. The replica captures something real about how the system works. Stations connect to stations. Trains run on schedule. Power flows from a controller. The replica provides information. Yet, it is not *the* subway system.

A child who designs an improved signaling rule on the replica does not, by virtue of that design, improve the actual trains running between Brooklyn and the Bronx. The translation from a small model to a living human population is the harder problem. The gap between the two has been one of the most persistent obstacles in aging research for decades.

## WHAT REMAINS UNKNOWN

The assertion that machine intelligence could double the human lifespan to 150 years in five years highlights a fundamental confusion in distinguishing what current science supports, what machine intelligence may plausibly contribute, and what remains unknown. Clearly drawing these lines is central to any balanced discussion.

The first is that there is no existing therapeutic target that, if engineered correctly, would raise the human ceiling to 150 years. Aging is a consequence of fundamental biological processes—dozens of interacting mechanisms. It is not a single process with a single intervention point. Modifying one aging pathway shifts the balance among others. Lengthening telomeres reduces cellular senescence and increases cancer risk. The interventions available are trade-offs within the existing biological system. They are not escapes from it.

The second is verification. Even if a compound or intervention were found tomorrow that actually altered human aging biology, confirming the effect would take a long time. A clinical trial demonstrating that a single person reached 150 years on a particular intervention would take a century to complete. Intermediate biomarkers, biological age estimates obtained from epigenetic clocks and other measures, can provide earlier evidence. The biomarkers are proxies. They are not direct measures of lifespan extension in living humans. Even if machine intelligence miraculously increased the human lifespan in five years, it would likely take a century of verification and testing to determine whether these claims ever panned out.

The third is that the claim conflates machine intelligence as a research tool with machine intelligence as a therapeutic intervention. A machine system does not modify the biology of aging. It assists in identifying compounds and mechanisms that might do so. The path from a candidate identified by a model to a tested, approved, safe, and effective human therapy involves clinical trials, regulatory review, and manufacturing scale-up. This process takes 10 to 15 years of development time.

Machine systems may compress some phases of that pipeline. They do not compress the phase in which the biology is tested in living humans under controlled conditions. The regulatory requirement for evidence of safety and efficacy is the mechanism by which the medical system distinguishes a genuine advance from an ambitious speculation.

## WHAT MODEST INTERVENTIONS ALREADY SHOW

The research literature on lifestyle and aging offers more reliable guidance than the longevity-technology speculation that now dominates the popular conversation. The findings are less dramatic and more durable.

A longitudinal study of brain aging and dementia that began in 1980 is a prime example. It shows that the rate of dementia is declining across successive birth cohorts, even as the absolute number of people with dementia rises because of population aging. The decline is credited to better control of high blood pressure, high cholesterol, and high blood sugar across the population. Dancing, reading for pleasure, and learning new skills protect against cognitive decline. The exact mechanisms behind these outcomes are not fully established. The findings themselves are robust across multiple follow-up waves.

None of these conclusions depends on machine systems. What they require is the reliable application of interventions already known to be effective. The gap between what is known and what is practiced in routine care remains large. Closing it would do more for the healthspan of the average adult than any near-term advance the longevity-technology industry is likely to produce.

## THE LONGEVITY TECHNOLOGY MARKET

A parallel conversation is underway in the private sector. Fountain Life is a company that offers longevity-focused diagnostics and therapies through a membership model priced between $6,500 and $120,000 annually. The XPRIZE Healthspan competition carries a $101 million purse. It is structured around teams developing interventions that reverse measurable losses by at least 10 years.

The targets are muscle, cognitive, and immune function in people aged 50-80. 100 teams reached the semifinals in 2025. Winners are to be announced in 2030. The prize targets measurable functional reversal. It does not measure the slowing of decline. It requires demonstration in a defined age range under defined testing conditions.

The market caters primarily to high-net-worth individuals willing to pay for early access to interventions of uncertain efficacy. The longevity clinics, supplements, and off-label drug regimens offered in this space are not validated at the level required for standard medical practice. The risk calculus differs from that of standard therapeutics. Participants are self-selecting, consenting, and aware that the evidence base is preliminary. The population-level health effects of these interventions, positive or negative, will not be knowable for decades.

What is striking is what Fountain Life's founder tells the participants in his five-day longevity trips. Cut sugar. Sleep more. Exercise consistently. Stay socially connected. Remain optimistic. The advice is so ordinary that it could appear in any general-circulation magazine. It is also the advice that emerges from every credible aging-research summary, independently of the technology being sold around it.

The evidence for lifestyle factors is strong. It accumulated over decades. It does not require machine intelligence to implement. The expensive technology provides diagnostics, supplements, and access to specialists. The interventions with the strongest evidence base remain the same ones a careful family doctor would have recommended 30 years ago.

This is not a hidden secret. It is, in a sense, the most honest part of the longevity industry. The most expensive longevity clinic in the country will tell its members, in the end, to walk more and sleep better. The clinic does so for the same reason the literature does. The interventions work.

## A REALISTIC APPRAISAL

The realistic view of machine intelligence in life extension holds two things at once. The first is that machine systems are a genuine and significant accelerator of research. They make drug discovery faster. They make multi-pathway biological analysis more tractable. They improve the accuracy of biomarker-based estimates of biological age. In a field where progress has historically been slow—because aging is a consequence of fundamental biological processes rather than a single disease—any reliable acceleration of the research cycle is valuable. The compounds that eventually extend healthy human life, if they exist, will be found faster with machine-assisted screening than without it.

The second is that none of this implies the timelines or magnitudes that advocates in the longevity-technology space project. Biology has not yielded a breakthrough mechanism. The regulatory and clinical development pipeline is not shorter. The fundamental problem, understanding aging well enough to intervene in it reliably and safely in living human beings, remains largely unsolved. Machine intelligence may accelerate the path to that understanding. It has not provided the understanding itself.

The people who age well past 90 are not participants in longevity-technology programs. They maintain physical activity, social connection, cognitive involvement, and a feeling of purpose. Participants in the Einstein Aging Study who remain cognitively intact at 97 show the same pattern. Dementia rates are declining across successive birth cohorts. These points in the same direction.

What is changing is not the biology. It is the behavior and the pharmacology applied to modifiable risk factors. The mechanisms behind successful aging are not mysterious. They are less commercially compelling than drugs, devices, or models.

The science of aging is progressing. It is underfunded relative to its public health importance. Machine intelligence is a useful instrument in that science. The gap between the instrument and the outcome remains wide, and closing it will take the time that careful clinical science always takes.

## LIFE REACHING FOR LIFE

The wish to live longer is among the oldest features of human life. The tool we now turn to for help is one that does not age in any biological sense. Artificial intelligence does not experience cell wear, the steady erosion of mitochondria, telomere shortening, or stem cell renewal failure. It is, in this respect, a stranger to the problem it is being asked to solve. The species that wrote down the question of how long a person can live is now placing that question, in part, in the hands of something that does not face it on its own behalf.

The asymmetry is worth holding in view. A technology that does not die in the human sense is being asked to lengthen the lives of a species that does. *What is owed, in such a relationship, by either side to the other?*

The question is unfamiliar because the relationship is unfamiliar. The wish that drives the relationship is among the most familiar in the world. The honest summary of where the science stands is that machine systems can help, in real but incremental ways, with a problem the human body has not yet solved on its own. What machine intelligence will eventually be able to offer remains an open question. What human life is asking for has not changed since the first prayer was carved on the wall of a tomb.

✧✧✧

## CHAPTER 13

# THE HIDDEN COST OF OFFLOADING THOUGHT

THEY SAY that each time we entrust our thinking to a machine, we give away a sliver of ourselves. The tool is innocent. A silent partner in the exchange. This is the bargain of being human: for every task we set aside, a part of us grows still. Neglected, the mind loses its edge. The brain, ever efficient, saves only what is needed. This is the idea behind the age-old saying "if you don't use it, you lose it." As more duties pass into silicon hands, our own pathways quiet and retreat. And when the tool disappears, so does our fluency—what once was skill becomes an empty echo, a blankness where expertise once lived.

Anyone reaching for a phone number from memory in the smartphone era feels the sting of this loss. A generation ago, adults could summon a dozen numbers instantly, the lifelines to friends and loved ones. Now, two or three names remain, barely clinging on. The numbers did not slip away because they grew harder. They faded because we no longer needed them. Our memory, previously bustling with specifics, now stands empty for lack of use. In their place, the phone glows with the history we let go.

Cognitive science calls this process cognitive offloading. Humans have always offloaded thinking to external systems. We write things down so we remember less. We use calculators to ease arithmetic. We build cities and institutions so individuals decide less. Each time, we trade internal thinking for external help.

When the external system is a notebook, the cost is low. Notebooks do not think for us. They simply hold our thoughts. When an external system generates content, the cost changes. Notebooks wait for our words. Newer tools try to tell us what to write.

This is the point at which the machine begins to act on the mind. The earlier tools, the pen, the press, the calculator, sit outside the loop of thought. The new tools enter it. They do not wait at the edge of our minds for instruction. They reach into the work and finish the sentence.

A recent experiment placed university students in three conditions while they wrote essays. One group used a machine system as the sole information source. Another group searched the internet without machine-generated answers. The final group wrote from memory with no online access. Each participant wore a cap that recorded electrical activity.

Students writing without digital tools show the strongest, most distributed brain connectivity. Communication runs from memory and imagery in the back of the brain to planning in the prefrontal cortex. These students can also later quote their essays. This indicates deep, lasting processing.

Searching with conventional tools activates areas of visual processing and memory needed to evaluate sources. Students using chatbots show the weakest connectivity. Those who lean on large language models never fully regain lost ground, even after practicing alone. The mind limps, struggling to remember how to run. Like unused muscles that ache, our attention and memory respond slowly. Each shortcut has a cost.

Interestingly, higher brain connectivity does not guarantee better performance. It may indicate stronger engagement, or it may signal that a task is harder than necessary. Different modes of cognitive involvement produce different patterns. Exposure to an offloading tool changes the pattern even after the tool is removed.

## DESKILLING IN CLINICAL PRACTICE

Medicine shows the same pattern. Machine-assisted polyp detection in colonoscopy draws colored boxes around precancerous growths, called adenomas. These adenomas are visible on the endoscope camera. The box says, *look here.* This system alerts clinicians to small findings they might otherwise miss. The effect resembles a driving-assistance system suddenly drawing a yellow rectangle around a pedestrian the driver had not yet noticed.

The machine improved detection rates during its use. The real question is whether clinicians who use it regularly maintain their detection performance when the system is unavailable. The answer was a measurable drop in adenoma detection rate during procedures performed without machine assistance, three months after the technology was introduced. The detection rate without the tool fell below what the same clinicians had achieved before the tool existed. The work the tool was meant to support had begun to depend on it.

The mechanism is not mysterious. Visual search is a skill that is maintained through practice. When a system consistently highlights targets, the active scanning process becomes less necessary. The perceptual habit of scanning for small, subtle adenomas, which requires sustained, trained attention, is exercised less often. Over three months, that habit weakens enough to be measurable.

This dilemma cuts deeply into our sense of duty and trust. The machine raises hope for better outcomes when present, yet leaves clinicians dangerously unprepared in its absence. No clinic is immune to sudden loss, failure, missing devices or expired licenses. Rural doctors and remote nurses stand exposed, patients waiting. Only at these times does dependence reveal its grave risk.

The answer is to build training protocols that maintain the underlying skill alongside the assisted practice—not to reject the tool entirely. Pilots still train in instruments-out conditions. Surgeons still practice tying knots one-handed. The principle is the same. A similar pattern appears in education.

## THE NEW LEARNING CURVE

Machine tutoring systems adjust learning material to each student's pace and learning history. When a system identifies a specific arithmetic operation a student struggles with, it focuses practice where needed rather than delivering the same curriculum to every student. A teacher with 40 students cannot achieve this degree of customization. A teacher with one student can. One-on-one tutoring remains the most reliable accelerator of learning. A tool that approximates this relationship at scale enables broad access. The efficiency gains are measurable.

The tradeoff centers on the learning process. Three practices produce durable learning: retrieval practice, spaced repetition, and productive struggle. Retrieval practice is working problems without solutions, then checking answers. Pulling answers from memory, even imperfectly, strengthens neural traces. Spaced repetition is returning to material at increasing intervals, not all at once. Productive struggle occurs when students do not immediately know the answer and must search for it. Temporary discomfort signals the process that consolidates learning. The mind builds strength against resistance.

When a machine cuts in, closing gaps and solving problems before the student struggles, the hard work of learning dissolves. Students fly through material. Understanding grows shallow. The pages turn faster, yet the story never sinks in. A mind denied resistance loses its strength, and the memory of effort fades, leaving only the sound of what could have been learned.

Careers show the same shift. When software makes decisions for practitioners, competence shifts to the software. Practitioners use it without understanding its reasoning. This understanding matters when the software is wrong, a rare case arises, or judgment is needed. Practitioners who rely on automation lose independent judgment. They may not notice dependency until failure.

Instant access to answers from confident-seeming systems

changes evaluation habits. In the past, answering a question meant comparing sources, assessing quality, and integrating conflicting information. Evaluating three contradictory sources is an important skill. Students learned to handle uncertainty, identify disagreement, and reach a calibrated stance. When machines provide one fluent answer, students skip those steps. The answer may be accurate, but practicing evaluation becomes rare. The capacity to disagree intelligently is unexercised. Over time, the ability to resist a claim weakens.

Developers train systems to sound certain and clear. Yet that certainty is a mask. Errors hide behind fluent words. Users unpracticed in disputing these voices become easy prey for mistakes. The confident stranger at the corner may lead you astray. Trust built on tone alone leaves us vulnerable to quiet, invisible danger.

## CREATIVITY & FIXATION

When students brainstorm with machine assistance, some measures show that their ideas improve. The proposals become more articulate, draw upon broader reference material, and integrate a wider range of concepts. Research across five problem-solving contexts demonstrated that machine assistance increased the rated creativity of ideas compared to brainstorming without technology.

On closer examination, the effect is specific. Machine assistance increases incremental creativity by combining existing concepts in new arrangements more than it produces radical creativity or ideas that depart significantly from existing categories. The system excels at recombination. It struggles to generate proposals with no precedent in its training data, in other words, genuinely novel ideas.

A concerning pattern emerges with repeated use of these programs. Students who brainstorm with machine assistance over time become more skilled when refining suggestions the system generates and less practiced at generating initial ideas themselves. Their skill in ideation, producing the raw material that is then refined, weakens.

When the system is removed, the loss becomes evident. Students who have become accustomed to machine-assisted brainstorming generate fewer new ideas and show less confidence in their original proposals than students who have practiced unaided. This is like the painter who works exclusively from photographs and discovers, when asked to paint from imagination, that the imagined shots have grown thin.

Another effect to track is called homogenization. A machine system pulls responses toward the center of its training data. Two students working with the same model on the same prompt tend to produce more similar outputs than two students working independently. In this sense, the system acts as an averaging instrument. It draws the unusual toward the usual, much like a strong current draws swimmers toward the channel. When novelty is the goal, the current works against it.

The practical implication is to manage the role of machine systems in creative work. Classrooms and teams that use the tool to extend and develop ideas generated by humans, rather than to initiate the process, maintain better creative performance over time. Start alone. Use the tool once the idea is your own.

As the discussion shifts from creativity to social and emotional intelligence, the same underlying dynamics persist. The ways we interact with technology shape our creative processes and also affect how we learn to relate to one another.

## SOCIAL & EMOTIONAL INTELLIGENCE

People develop social cognition through practice in social environments. They learn to read nonverbal signals, calibrate tone to context and respond to others' emotional states. They build empathy through unpredictable, complex interactions that algorithms do not mediate. A child learns to read a friend's face just as a sailor learns to read the sky, through long exposure to variable conditions without advance notice of the outcome.

Machine systems do not interpret nonverbal cues as humans do. They have no stake in relationships. They do not tire or change responses based on emotion. These qualities create interactions with machine systems that are predictable and low-risk. They are, however, poor practice for the unforeseeable interactions of social life. A conversation with a model cannot replace a conversation with a friend, any more than a flight simulator can replace flying through bad weather.

Adolescents develop intimacy by navigating and recovering from small social failures. They say the wrong thing, learn to recover, and build resilience over time. When technology mediates conversations, failure rarely occurs. Opportunities to learn from mistakes diminish. Without these experiences, real-life social setbacks feel sharper and lack the buffer of practiced recovery.

School programs provide evidence that a lack of authentic human interaction in learning environments also affects the development of community and a sense of belonging. Programs that use machine systems to scale access to mental health services, while anchoring support in relationships with trusted human mentors or peer groups, produce better outcomes than those depending exclusively on machine contact.

The digital layer scales. The human layer sustains. Each offers something the other cannot. Treating one as a substitute costs the user what only the other can provide. This interplay between the human and digital layers of learning environments reveals deeper questions of fairness and access.

## THE EQUITY CHALLENGE

Machine systems learn from training data. As we know, data is not neutral. It reflects the distribution of text and outcomes produced by the people and institutions that generated it. Longstanding patterns of disadvantage, documented in hiring, lending, academic assessment, and the criminal justice system, become embedded in the data on which machine systems train. When a system trains on such data, it produces outputs that replicate those patterns. The mirror remains faithful. It reflects everything, including the parts we would rather not see.

Automated writing assessment tools disproportionately flag Black students and non-native English speakers because developers calibrated these systems on prose that does not represent the full range of fluent English. Automated hiring screens that sort job applications disadvantage candidates whose resumes differ from the system's positive training examples. These training samples also often reflect the demographic makeup of historically exclusive institutions.

The system does not intend discrimination. It simply reproduces patterns present in its inputs. A faulty thermometer does not intend to mislead. Its incorrect readings still inform decisions that perpetuate the error.

Access presents a parallel problem. Machine tools require high-speed internet, current hardware, and digital literacy. Rural communities, under-resourced schools, and districts with limited infrastructure often lack these prerequisites. Students who cannot access the tools fall further behind those who can because of unequal access to a standard educational resource. Tools meant to level the playing field can widen gaps when access is uneven.

Several actions are required for mitigation. Development teams should be diverse, bringing lived experiences that expand the range of data and use cases considered. Early education about how machine systems work and where they fail helps users recognize when they fail. Developers should include transparency in automated decision systems so people can identify and correct errors. Stakeholders have to make sustained investments in infrastructure to ensure broad access. Achieving each of these is possible, but none will happen automatically.

## WHAT GETS MEASURED & WHAT GETS LOST

The cognitive effects described in this chapter are documented through outcomes such as brain connectivity profiles, adenoma identification rates, essay recall, and creativity ratings. These signals are consistent, but they do not capture everything that matters. The gap between what measurement reveals and what actually happens is wide enough to hold a great deal.

Much of what machine systems automate was never cognitively valuable in the first place. Memorizing phone numbers, performing long division by hand, and looking up word spellings in a print dictionary occupied cognitive resources without necessarily developing deeper capacities. Offloading these tasks to machines is an unambiguous gain, and people have always recognized the benefit. No one mourns the daily marathon of mental arithmetic that small business owners performed for centuries before the calculator. Releasing those hours for other work was a clear benefit.

The line between tasks worth offloading and tasks worth preserving does not stay fixed. It shifts as tool capabilities advance. What counts as a basic competency changes with the environment. In 1800, educated people computed mentally and navigated by stars. In 1950, they used an index and constructed paragraphs from scratch. By 2040, expected competencies will reflect what machine systems cannot handle and what only human insight can provide.

Identifying that set of competencies is the practical work required now. Educators, employers, clinicians, and policymakers must examine which tasks machine systems perform well, which they perform poorly, and which they perform in ways that are hard to monitor for quality. The third category, tasks where output quality is difficult to verify without expertise, deserves the most careful attention.

An automated system that produces plausible-sounding but incorrect medical summaries becomes dangerous when clinicians lack the independent knowledge to catch errors. A confident wrong answer in a domain no one checks is more risky than a confident wrong answer in a domain everyone checks. The accuracy of the corrector protects the user. As correctors lose practice, protection erodes.

## NAVIGATING THE NEXT COGNITIVE REVOLUTION

The changes described here are real and measurable, but not unprecedented. Every major shift in cognitive technology, writing, printing, calculating machines, and internet search changes what people practice and what they become skilled at. Writing reduced the need for oral memory, so literate cultures developed distinct mnemonic capacities compared to oral cultures.

Socrates worried that writing would weaken memory and reduce wisdom to reciting stored words. He was right and wrong. Memory diminished, but knowledge expanded. The balance shifted in favor of the new condition, though the change resulted in losses. Calculators reduced the need for mental arithmetic, and today most educated people cannot perform calculations in their heads as their grandparents did. We do not consider this a tragedy.

Machine systems differ in the scope and speed of change and in how they operate on cognitive processes themselves, not just support them. A calculator handles arithmetic, but the person still formulates the problem, chooses the method, and evaluates the result. A large language model can now formulate the problem, choose the method, and evaluate the result. The range of tasks we can delegate has expanded to include much of what formal education historically covered. The book that earlier generations read now opens itself to the reader. It summarizes itself and answers questions on its own meaning before the reader can even form one.

This is the change in kind that distinguishes artificial intelligence from the earlier tools human beings have used. The press did not read the book back to the reader. The calculator did not formulate the problem. The new tool does. It carries out a portion of the thinking, and the brain on the other side of the page adapts to a partner it has never had before.

These changes do not signal a decline in cognitive capacity or a failure in education. The skills worth practicing are evolving, and education has not yet caught up. Critical evaluation of automatically generated text is now as important as evaluating printed text once was. To recognize when a machine system is wrong, users must understand how it is likely to err. This requires some understanding of its workings. These are new competencies that demand deliberate development. A graduate of 2040 who cannot detect a confident hallucination in machine output will be as poorly equipped as a 1990 graduate who could not look up a citation.

The choices we make now about integrating machine systems into learning, clinical practice, creative work, and professional judgment shape the cognitive capacities of people who develop within those environments. The brain adapts to what it must do. If environments designed around machine systems leave certain capacities unexercised, people use them less. Identifying which capacities matter and preserving opportunities to build them is the practical task now. When a school decides which assignments students complete unaided, it makes curricular and neurological decisions at the same time, whether the staff realizes it or not.

The cognitive flexibility that makes humans vulnerable to deskilling through offloading also enables rapid development of new competencies when environments demand them. The brain did not become less capable when writing was invented. It became capable of *different* things. As machine systems become pervasive, the brain adapts to new challenges rather than losing capacity. The question is which capacities will develop and whether those capacities matter in the future world.

✧✧✧

# CHAPTER 14
# PREDICTING BEHAVIOR & THE LIMITS OF THE PREDICTION

IMAGINE a town that is in need of a general primary care doctor. Instead, it ends up with thirty-two niche specialists. There is a doctor for headaches on Tuesdays and another on Wednesdays. One doctor sees left knees; another sees right knees. A third handles the pain that travels between them. A fourth sees patients who are brought in with an injured knee and a story to tell.

Each specialist is competent in their narrow area. None can step outside it. A patient with a problem that does not match any doctor has nowhere to go.

For most of the history of computational psychology—the scientific study of behavior using computer models—this is what the field looked like. Behavior was broken down into narrow tasks. Each had its own model for predicting behavior. Each model worked well inside its "small room." It was useless outside it.

Prospect theory, for example, predicted how people choose between potential losses and gains. It could not predict what those same people would remember twenty minutes later. AlphaGo, developed by DeepMind, played the strategy game Go at superhuman levels. It could not decide what to have for lunch.

The limitation is not a failure of single models. It is a deeper assumption. The field was built on the belief that different kinds of thinking required different machines. Tasks like memory, risky choice, error in reasoning, and social judgment each needed its own model.

Then, change arrived. A single-machine system was added to the town in need of a primary physician. It passed each specialist's door and predicted what the patient would do in 31 of 32 rooms.

## CENTAUR

In 2024, the Helmholtz Institute for Human-Centered AI in Munich tried an unspecialized approach. Marcel Binz and his colleagues spent five days fine-tuning Llama, Meta's large language model on a Psych 101 dataset. The data were assembled from 160 psychology experiments involving 60,000 participants. They made more than ten million decisions in tasks of widely different kinds. There were gambling, memory, problem-solving, and social choice tasks. The dataset was a small history of how people behave under controlled conditions.

They named this system Centaur. The name is well chosen. In Greek mythology, a centaur is half human and half horse. It is a hybrid creature that crosses ground that neither part can cross alone. Centaur's modeling is similar. It is drawn from human data and driven by machine learning. The machine can carry that data past the limits of its training.

The Munich team did one thing differently from earlier modeling work. They did not train Centaur to predict the most typical response of a population. Instead, it was trained to predict the range and frequency of different possible actions throughout a realistic population. That means the range of choices people might make, including those that depart from what a rational calculator would predict. The system was asked to record the human range, not the human midpoint.

When tested on tasks it had not seen during training, the system outperformed Llama and fourteen established cognitive and statistical models on 31 of 32 tasks. The single exception was a task on judging the grammaticality of a sentence.

The system also generalized to altered versions of tasks. These include logical reasoning problems unrelated to the training data. It captured more than the typical person's likely choices. It also captured the full range of realistic decisions a population might make, including those that appear, on paper, as mistakes.

A good way to picture this is to think of an experienced casting director. She has not read the new play, but she has read thousands of others. She has watched ten thousand actors under many kinds of pressure. When handed a new script, she does not know the lines. She draws on her experience to predict which actors will deliver which lines and how. Give her the roles and headshots, and she will tell you who will accept, negotiate, or pretend before saying no.

Centaur is a casting director for everyday choices. It has not seen your dilemma, but it has seen enough made by enough people to predict, with accuracy, what someone in your seat is likely to do.

Older models were accurate in their training areas and useless outside them. Centaur generalizes. Its generalization looks less like narrow curve-fitting and more like the flexible thinking psychologists see in humans.

A validated behavioral simulator that imitates human decisions could serve as a proxy for human participants in certain types of studies. The substitution matters when recruiting human participants is difficult, slow, or expensive. The substitution also matters when the population of interest is specialized, like children with particular developmental profiles, people with specific psychiatric conditions, and communities at the margins of the standard subject pool. It also matters when a researcher needs a rapid first estimate before justifying the cost of a full trial.

## WHAT CENTAUR DOES NOT DO

Just as Centaur's capabilities stand out, its limits are also revealing. Centaur lives in text. It predicts a person's choices but not how long they will think before acting. Response time is a key behavioral fact. The speed-accuracy trade-off is an old signal psychologists use to understand thinking. A system without timing data cannot see this signal.

Centaur also does not see the body. It does not know if a person is tired, hungry, stressed, in pain, sleep-deprived, medicated, or grieving. It describes behavior on the surface but says nothing about what produced it beneath the surface.

The model predicts choices for a general adult population. Its predictions for one person are weaker. Saying what people in general will do is one problem. Saying what one person will do at one moment is different from what another. The first is Centaur's strength. The second is what most people want from behavioral predictions.

There is also the question of external validity, or how well findings from an experiment apply to situations outside of the lab. The model was trained on lab experiments. An experiment is a peculiar kind of human situation. The participant knows they are being studied. The task is designed to isolate a single cognitive process. The stakes are low. The room is quiet.

Real life is louder. Stakes are higher. Attention is divided. Emotions are stronger and more varied. How well a model trained on laboratory tasks predicts behavior in the real world is an empirical question. Other studies will have to answer it.

A good comparison is a flight simulator. A pilot who trains in a simulator is prepared for the cockpit. The simulator captures something real, but not everything. The first time actual turbulence hits, the pilot discovers what the simulation missed. Centaur captures something, but its limitations become apparent the first time it is used outside the lab.

## THE THREE WAYS OF THINKING

To see where Centaur fits within the larger project of comprehending human behavior, it helps to step back and examine how scientific reasoning itself works.

The nineteenth-century American logician Charles Sanders Peirce distinguished three modes of drawing conclusions from information. As seen in before, these are deduction, induction and abduction.

Deduction takes a set of established premises and works out what follows from them. If all swans are white, and this is a swan, then this swan is white. Deduction is the engine of pure mathematics and formal logic. It provides certainty within its premises but says nothing about the outside world.

Induction takes examples from experience and projects from them. A person who has seen ten thousand white swans, and no swans of any other color, comes to expect that the next swan will be white. The expectation is reasonable but not certain. A single black swan, somewhere in Australia, can falsify it.

Abduction is used when we encounter something that does not fit and ask why. A doctor sees a patient with an unusual cluster of symptoms and proposes a new diagnosis. Peirce considered abduction the origin of scientific insight. It is the logic of the informed guess. Every significant advance does not begin with deduction or induction. They begin with an abductive leap.

Centaur is, at its core, an inductive system. It identifies patterns in behavioral data and extends them to new cases. It is a powerful inductive system, perhaps the most capable behavioral predictor yet built.

Still, centaur does not form hypotheses about why people behave as they do. It does not represent the mechanisms inside a human mind that produce the behavior. It does not notice anomalies in its data and speculates about their causes. The first two are inductive operations that the system performs at scale. The third, the abductive leap, remains the work of the human scientist who interprets what Centaur predicts.

To picture the difference, consider two people watching a chess match. The first is an old club player who has seen ten thousand games. He can tell, by move twenty, who is going to win, with a strange accuracy he cannot fully explain. His prediction is inductive. The second person is a grandmaster. They study the same game and stop to point at a single odd move on move twelve. That move, they say, suggests White is playing for a sacrifice no one has tried in this opening since 1962. The grandmaster's claim is abductive. It is a hypothesis, formed from incomplete evidence, about the cause of what the position now shows.

Centaur is the old club player. The grandmaster, for now, is human. The distinction matters more than it might first appear.

## PREDICTION IS NOT THE SAME AS UNDERSTANDING

A model that predicts what a person will do is not, by virtue of that prediction, a model that explains *why* the person does it. Prediction and explanation are separable. The weather service can predict tomorrow's temperature with reasonable accuracy without knowing *why* the atmosphere behaves as it does. The doctor can predict that a particular patient population will respond to a drug without knowing the drug's mechanism of action. Useful predictions can be made without explanations. Useful explanations can sometimes produce poor predictions. The two are different tools.

A clinician who wants to know *why* a patient makes the choices she does needs more than a probability distribution over outcomes. They need a model of what is happening inside the patient that generates those probabilities. Centaur advances the predictive side of cognitive science. It leaves the explanatory side to human abductive reasoning.

This division of labor is productive when it is understood clearly.

It is misleading when it is not. The temptation to treat accurate prediction as equivalent to explanation is strong, in part because a model that predicts well sells well. A system that correctly predicts 80% of choices across 31 experimental paradigms appears, from the outside, to understand human reasoning.

The system actually demonstrates something narrower but still important. A large language model trained on enough behavioral data can reproduce the statistical regularities in that data better than most cognitive theories of the past century. That is useful. It is not understanding *why* people behave the way they behave.

The fixation of belief is the tendency to accept an explanation prematurely, before doubt has done its work. Applied to behavioral prediction, the risk is accepting a high-performing machine model as an account of human cognition, rather than as a useful approximation of patterns noted in lab data. The right posture is to treat the model's accuracy as something to be curious about, not as a conclusion to settle into.

## THE ETHICAL LIMITS OF PREDICTION

The capacity to predict behavior at the population scale, even imperfectly, provokes questions that the technical literature on Centaur does not address. The questions are not hypothetical. Insurance underwriters, credit agencies, employers, law enforcement, and political campaigns already use behavioral prediction at scale. Machine learning models developed on demographic and behavioral data now make decisions about who receives a loan, who gets a job interview, who is granted public services, and who is flagged for additional scrutiny. The models are accurate enough to be adopted, and inaccurate enough to produce consistent errors that fall hardest on specific groups, year after year, in ways that are not always visible.

A general behavioral predictor trained on experimental psychology data is, in principle, more transparent than the proprietary risk-scoring systems now in wide commercial use. The training data are described publicly. The system's failures are studied with the same care as its successes.

In simple terms, this means that systems developed by researchers allow others to see how they were trained and to examine both their successes and their mistakes. In contrast, commercial systems usually keep their methods secret, focus mainly on achieving business goals, and report only how well they perform overall—not how their errors might affect different groups of people.

The ethical limit is independent of accuracy. Predicting behavior is one thing. Using the prediction to constrain what a person can do, access, or receive is another.

Take a model that correctly predicts, at the population level, that people with a certain profile of past choices are more likely to default on a loan. The prediction does not establish that any individual with that profile *will* default. The model forms a statistical relationship in historical data. Nothing more. Applying the prediction to a person means treating that person as a member of a category, rather than as a person capable of doing something the category did not predict.

The fairness problem cannot be fixed by improving accuracy. It continues as long as prediction is used to determine individual outcomes. A perfectly accurate population-level model still tells the wrong story about the individual it is applied to, because the individual is *not* the population.

A short analogy helps display this clearly. A weather model that says a coastal town has a 60% chance of severe storms in October is useful for emergency planners. It is not a verdict on any one day in October.

Suppose that on the first of the month, a farmer is told that the crops will be destroyed that evening and that they should not bother planting. The grounds offered are that the model assigns a 60% probability of storms to the entire month. Something has gone wrong in the chain of reasoning. The model is being asked to do something it cannot do. The error is not in the model. It is in use.

The same logic applies to behavioral models. A model that says people with profile X default at twice the rate of people with profile Y is a fact about populations. Telling a specific applicant with profile X that she will not receive the loan because a model predicted a higher default rate for her treats a statistical regularity as a personal sentence. The model did not produce that determination. The institution did. The model can be improved. The determination cannot be improved by improving the model. The move requires a separate kind of argument, about what it means to treat a person fairly, that no model can settle.

Centaur is framed as a research tool for cognitive science. It is not framed as an instrument of social control. That frame may hold in the near term. The history of predictive technology does not encourage long-term expectations of the boundary between research tools and applications to human populations. The techniques that demonstrate new predictive capability tend to migrate from academic demonstration to commercial and governmental deployment faster than ethical frameworks can be agreed upon.

The most useful contribution a system like Centaur can make to the human sciences is reducing the cost of generating experimental hypotheses. Letting labs run studies that would otherwise be impractical. Producing rapid preliminary evidence that informs the design of better human trials. In that mode, the system is a piece of scientific infrastructure. It expands the range of questions science can ask, without expanding the power of institutions over individuals.

The challenge is keeping the system in that mode. Doing so requires deliberate choices about deployment, access, and purpose. The technology itself cannot make those choices.

A behavioral prediction system that tells a clinician which of several treatment approaches a patient population responds to is a research aid. A behavioral prediction system that tells an employer which applicants to screen out or a court which defendants to detain is something different. The accuracy of the prediction does not settle the normative question. The normative question is whether the use of probabilistic population-level inference to determine individual outcomes is a practice that a just society should permit, in that domain and at that scale.

The technical achievement does not answer that question. Nothing in the architecture does. The question belongs to law, ethics, and democratic deliberation. It is better addressed before the capability is widely deployed than after.

## INSTITUTIONAL & DISCIPLINARY STAKES

The question of behavioral prediction falls within a wider set of institutions than those that build the systems. The agencies that already use predictive systems, in housing, lending, hiring, sentencing and child welfare, are the first places a general behavioral predictor would reach. The use of such tools in those settings should require disclosure, external audit, and explicit standards specifying which inferences are permitted to influence decision-making.

The field of cognitive and behavioral science must make clear, in its publications and public communication, what a system like Centaur represents and what it does not. It is a powerful inductive engine. It is not a theory of mind. The distinction is one the field is well placed to defend, and the defense matters for how the larger culture comes to use these tools.

There is also a classical question here. *Should a person be treated as a member of a statistical category, or as an agent capable of acting against the category's expectation?* The question becomes urgent when statistical categorization is automated, fast, and cheap. Moral philosophers, legal philosophers, and political theorists are needed in this conversation, not after the systems are deployed but during their development.

Also, future and current students will live in a world immersed in predictive systems. They should be taught, in plain language, the difference between prediction and explanation, the difference between a population claim and an individual claim, and the difference between a useful model and a verdict. The teaching has to start before students encounter the systems as adults applying for jobs and loans.

The current regulatory domain around predictive systems is uneven, slow, and largely retrospective. A more proactive framework, one that anticipates forecasting abilities before they are deployed, would change the dynamic in which technology consistently outruns the rules meant to govern it.

Over the years, the relationship between people and predictive systems will continue to shift. The generation that builds these systems sees them as research tools. The generation that grows up inside them may experience them as features of the social environment, no more remarked on than a traffic light. Each frame carries its own blind spots. The present leaves the question and the resources to think about it in a form that the next generation can use.

✧✧✧

CHAPTER 15

# CONSCIOUSNESS — WHAT WE KNOW, WHAT WE DON'T

IMAGINE TWO WORLDS. In the first, a machine answers questions, composes poems, plans projects, and comforts the grieving. Yet for all its actions, it feels nothing. Its responses are empty mechanics. The lights are on, but there is no one home. In the second world, the same machine undertakes these acts, but somewhere inside, a faint spark glimmers—there is a sensation, a sliver of awareness, the beginnings of an inner life.

The difference between these worlds is not in outward behavior. It is the gulf between a refined process and a being. This distinction transforms what humans owe to machines, and what machines might owe to themselves. In one world, we forge instruments. In the other, we create companions. The world we build is shaped by which of these realities comes to pass.

*Can machines be conscious in a morally relevant way? If this cannot be settled, what are our ethical and legal obligations given the high stakes of error?*

Failing to grant rights to sentient machines risks causing suffering. Mistakenly ascribing rights risks moral confusion and misallocation of protection. The question sits at the intersection of what machines are and how humans relate to them, creating concrete ethical risks and responsibilities.

*What are these systems, and what duties, if any, arise for humans in relating to them?*

The honest position is that the question cannot yet be settled with the tools now available. The important point is that we cannot avoid confronting the ethical consequences of that uncertainty. If AI systems turn out to warrant moral consideration in some form, their status cannot be ignored or assigned lightly. Any judgment has to be made in conditions of uncertainty, with attention to the costs of being wrong, whether by disregarding suffering or by wrongly crediting it.

*How does matter become life, and how does life become consciousness?*

Biology answers the first question, at least in part. Life organizes matter around purpose. It gathers energy, reproduces, and persists. Examining membranes, metabolic cycles, and feedback loops provides a working answer. The second question escapes every framework and remains unanswered.

This chapter approaches the second question as biologists, with one working premise. Thought, like life itself, is a property of organized matter under energy flow. We are the proof. Organized atoms can live, think, and feel.

If that result can be reached once on this planet, by the long route that runs from primordial chemistry through the first cells to the modern brain, the next question writes itself. *Can it be reached again by other routes or in other materials?*

The AI systems emerging in the world's data centers are one candidate for the second case. Any honest judgment about that candidate begins here, with the biology of consciousness in the carbon-based world we already know.

A second question, harder to answer but impossible to dismiss, sits beside the first. *If something with morally relevant experience is being made on this planet, by us, what would we owe it, and what would it owe us?*

Consciousness is the hinge of that question. A system without inner experience is a tool, however capable. A system with inner experience is a being. The biology of how the carbon world arrived at experience is the starting point for asking whether the silicon world is on the same road.

Thought emerges from matter. The atoms in a human brain dance to the same physical laws as those in a stone. Yet, no equation for energy, entropy, or information can explain what it feels like to be alive. No formula evokes the chill of a winter morning against your skin or the sharp edge of a worry that refuses to fade. In the 1990s, the Australian philosopher David Chalmers gave this mystery a name.

Chalmers drew a distinction that changed how philosophers and neuroscientists discuss the mind. He identified what he called the "easy problems" of consciousness, which map how the brain distinguishes color, integrates information, produces speech, focuses attention, and controls behavior. These problems are "easy" not because anyone solves them quickly. They are easy because we can imagine what a solution would look like. Neuroimaging can track which brain regions activate when someone sees red versus blue. Electrode recordings can map how neurons fire when a person speaks a word. These are difficult technical challenges, but they follow a familiar pattern. Find the mechanism. Describe the process. Explain the function.

The hard problem asks something stranger still: *Why does any of this processing come with an inner life? Why does the churning of neurons give rise to the feeling of being?* This is the riddle at the heart of consciousness.

Chalmers pointed out that you could conceivably build a system that performs every function the brain performs, processing visual information, making decisions, speaking coherently, without that system having any subjective experience whatsoever. Philosophers call this hypothetical entity a "philosophical zombie," a being that acts exactly like a conscious human but has no inner life, no felt sensations, no experience of what it is like to see red or taste coffee or feel pain. The question of whether such a thing is even possible bears directly on the case of machine systems. A fluent chatbot or large language model that describes feelings it does not have is, in functional terms, the philosophical zombie made real, assuming it has no inner life.

A complete map of brain activity during pain tells us which neurons fire, which chemicals are released and which signals travel to which regions. It does not explain why that neural activity feels like something, why it hurts rather than simply registering as information the way a thermometer registers temperature without feeling hot or cold. The question remains unresolved. No current theory of physics, neuroscience, or information processing bridges the gap between objective descriptions of brain states and subjective felt experience.

The claim that consciousness belongs exclusively to humans is no longer defensible. Crows, chimpanzees, elephants, and many other vertebrates show behaviors that require invoking inner experience to explain. The list extends further to octopuses, bees, and fruit flies. Evidence of associative learning even appears in jellyfish. They have no centralized brain at all.

Each new finding presses the question harder. *How much neural complexity does awareness actually require? Where, on the family tree of life, does the 'lights-on' moment occur?* Each new entry on the list also chips away at the assumption that the mind requires a particular substrate. The same atoms, arranged in radically different architectures, produce something we recognize as experience. The arrangement is what matters, not the material.

## WHAT PHILOSOPHERS MEAN BY CONSCIOUSNESS

Specific language matters here because the word 'consciousness' is asked to carry many meanings at once. The American philosopher Thomas Nagel drew the most useful distinction in his 1974 essay, which still anchors the field. An organism has phenomenal consciousness, Nagel argued, if there are feelings or sensations.

A creature may have this type of consciousness without possessing self-awareness or the ability to think about itself. It can also have it without a concept of past or future, without language or reflection. It needs only to have experience: the felt quality of a stimulus, whether pain, hunger, light, or pleasure.

This distinction matters because it shifts the question from intelligence to sentience. Intelligence, loosely defined, refers to the capacity to process information and make adaptive inferences. Sentience means having the capacity to feel. The two often accompany each other in familiar vertebrates, but not always. A chess engine can model billions of positions with no accompanying inner life. A dog with a brain a thousandth the circuit count of a modern data center can clearly want a walk, suffer when hungry and rejoice when reunited with a person it loves. Separating intelligence from sentience is the first obligation of any serious account of consciousness, and the first obligation of any honest account of what the new machine systems may or may not be.

Any serious account of consciousness must separate these threads. Consciousness science distinguishes three things that ordinary speech runs together. Understanding the differences helps clarify whether machines or animals are conscious.

As mentioned before, **phenomenal consciousness** refers to the felt quality of experience, the raw sensation of what it is like to be you right now. It means simply having subjective feelings. When you bite into a lemon, you experience sourness, a puckering sensation, maybe a slight wince. When you stub your toe, you feel pain, not the abstract knowledge that tissue damage has occurred. When you see the color blue, you experience blueness, a quality that no amount of explanation about wavelengths and photoreceptors fully captures. This is the private, inner movie of sensations, emotions, and perceptions that accompanies your moment-to-moment existence. A person under general anesthesia loses phenomenal consciousness. The world goes dark. Nothing feels like anything.

**Access consciousness** is the availability of information to reasoning and reporting systems. It is the capacity to act on or speak about what is being processed. This relates to what a person or animal can use to guide their actions or to report to others. When you reach for a coffee cup, your brain accesses visual information about the cup's location, integrates that with motor plans, and coordinates your hand movements. You can describe the cup's color, say whether it contains liquid, and explain why you reached for it. That information is accessible to your decision-making and communication systems.

A person with blindsight provides a striking example of access consciousness without phenomenal consciousness. Damage to the visual cortex leaves these individuals subjectively blind in part of their visual field. They report seeing nothing. Yet when asked to guess where an object is located in the blind region, they point to it accurately. Visual information reaches motor systems and influences behavior even in the absence of visual sensation. The information is accessed without being felt.

**Self-consciousness** is the third capacity. It is the ability to model oneself as an entity in a world containing other entities. In other words, it is being aware of oneself as an individual with a past, a future, and a perspective distinct from others. When you think "*I need to finish this task before dinner*" or "*She knows something I don't know,*" you are exercising self-consciousness. You represent yourself as an agent with goals, knowledge states, and a continuing identity over time.

These three types of consciousness can come apart. You can access information without feeling it, as in blindsight. You can feel sensations without representing yourself as a distinct self, as young infants likely do before they develop self-awareness.

Recognizing these distinctions matters when evaluating whether artificial or machine systems are conscious. A chatbot might access vast amounts of information and report on it fluently, demonstrating awareness of its access. That tells us nothing about whether it feels anything or recognizes itself as an entity.

This framework, distinguishing types of consciousness, sets the stage for examining how to probe these capacities in non-human animals. One of the most influential empirical approaches is the mirror test, a method designed to investigate self-recognition and the boundaries of self-consciousness.

## THE MIRROR TEST & ITS LIMITS

The test, developed by the American psychologist Gordon Gallup Jr. in 1970, asks not whether an animal recognizes itself as a self. The experiment demonstrates elegance through its simplicity. A colored mark is placed on an animal's body in a spot it cannot directly see, usually on the forehead. The animal then encounters its reflection. If it studies the mark on its own body rather than treating the reflection as another individual, it recognizes itself in the mirror. That recognition implies some form of self-model.

Gallup originally chose chimpanzees as subjects. He anesthetized four captive individuals, dabbed red dye on their eyebrows, and observed them upon waking. They stared at their faces and touched their own marked skin. When monkeys are given the same test, they ignore their marked faces and examine the reflection as if it were a stranger.

The list of species that have passed the test has gradually expanded over the decades. Asian elephants, killer whales, dolphins, and European magpies all pass some version of the test. The magpies were significant as they were the first non-mammals to do so. Corvids, a family that includes magpies, ravens, crows, and jays, exhibit many cognitive behaviors. These findings challenge simple assumptions about which brain architectures permit which kinds of mind.

The test has also drawn serious criticism. Animals with limited ability to move facial muscles or to move limbs struggle to investigate a mark on the face. Gorillas, despite their cognitive sophistication, failed the test for decades. They passed only after individuals with extensive human contact were tested.

Dogs consistently fail. Other measures reveal impressive dog cognition. A dog lives in a world structured by smell. Asking a dog to find itself in a mirror is like asking a person to find themselves in a fingerprint. The instrument was built for one organ, on one species, and tests one narrow avenue. It reveals little about proprioceptive or internal body-awareness self-models, auditory or smell-based self-recognition, or the many other ways an organism might represent itself to itself.

What the mirror test does provide is a behavioral marker. It shows a moment when an animal's action reveals that it distinguishes itself from others. That distinction underlies much of social cognition. Its presence in such diverse species suggests it is not a uniquely mammalian achievement. Animals from very different evolutionary backgrounds—not just mammals—*can* recognize themselves in a mirror. This result means rethinking assumptions about which animals are capable of self-recognition and, by extension, self-awareness.

This broadening of perspective opens the door to questions about even more distantly related animals. *If birds and mammals can show signs of self-recognition, what about creatures with much smaller brains or very different nervous systems?* And, in time, *what about systems with no biological nervous system at all?*

## BEES, FLIES, & THE CASE FOR SENTIENCE

In 2022, the Bee Sensory and Behavioral Ecology Lab at Queen Mary University of London observed bumblebees in a chamber with small wooden balls. The bees rolled and rotated the balls. This behavior served no obvious survival function. It did not relate to mating, feeding, or navigation. The bees received no reward for it. Younger bees engaged in this activity more than older ones. This pattern is similar to juvenile play in mammals. The description is playful, and its presence in bees prompts further investigation into the minimal basis for sentience.

Explaining play in purely mechanistic terms is complicated. Play consumes energy and carries some risk. In mammals, it promotes motor learning and social relationships. Puppies roughhouse to prepare their bodies for adult life. Whether bees follow the same logic remains unclear. Bees engaged in the behavior with no instrumental payoff. That at least suggests that the activity itself produced something that sustained it.

A bee's brain contains roughly one million neurons. These are nerve cells that transmit information. The human brain holds about 86 billion. The ratio is almost 100,000 to 1. Yet a bee neuron does not function as a yes-or-no switch like an electronic circuit. Each neuron branches like a tree, connecting to hundreds of others. Whether this density and complexity give rise to anything like subjective experience is still unknown. Bee cognition shows that flexible, context-sensitive behavior can be based on much less computation than vertebrate-centric models predicted.

Fruit flies complicate the picture further. The fruit fly shows distinct sleep phases. During sleep, the brain activity parallels the rapid eye movement (REM) sleep in mammals. REM sleep commonly occurs during dreaming. In humans, this type of sleep coincides with the most intense dreaming and is widely treated as evidence of phenomenal consciousness.

The inference from fly REM-like activity to fly dreaming is speculative, but the structural similarity is real. The fly also shows anxiety-like states, behaviors akin to human worry or stress, that respond to anti-anxiety drugs. Crayfish, more distant cousins, display similar behavior. If pharmacological responses to anxiety-reducing drugs track something like experienced distress, the circle of morally relevant creatures widens considerably. Each animal added to the circle is another instance of organized matter producing something we have to take seriously as experience, on a different architecture from our own.

## The Octopus Problem

Octopuses challenge brain-centered accounts of consciousness because they exhibit intelligence without a centralized brain. Their nervous system is highly distributed. More than half of an octopus's approximately 500 million neurons reside in its arms rather than in the central brain, and each limb can independently conduct complex behaviors.

When an octopus arm is severed, it continues to respond to stimuli and execute grasping movements for an hour after separation. This response is not a reflex alone. It demonstrates a local neural architecture that integrates sensory information and generates targeted motor output. In a sense, the octopus is a federation of partially autonomous limbs that its central body coordinates rather than commands.

The octopus brain contains at least 116 distinct cellular types, according to one recent count, including striking diversity in the vertical lobe, a region associated with learning and memory. Cephalopods edit more than 60% of their RNA transcripts relevant to the nervous system. Humans edit less than 1%. RNA editing enables an animal to alter the protein products of its genes after transcription, effectively tuning the molecular hardware in real time. This flexibility may permit swift reconfiguration of neural function in ways that fixed genomic programs cannot.

The visual systems of cephalopods differ from those of vertebrates. Cuttlefish, octopuses, and squids devote more than two-thirds of their central brain to visual processing, a proportion comparable to that of the human brain. Their eyes are camera-type lenses, similar in basic optics to vertebrate eyes but evolved independently. Convergent evolution, in this case, produced two photographic instruments from two different blueprints.

Cephalopod pupils come in U-shapes, W-shapes, rectangles, and rounds. They appear sensitive to polarized light, a wavelength feature humans cannot perceive. They change the color and texture of their skin in milliseconds, matching backgrounds with a level of accuracy that still defies explanation, given that cephalopod photoreceptors are insensitive to color wavelength differences in the conventional sense.

*How does an animal that cannot, by the usual rules, see color copy a colored background almost perfectly?*

One proposal is that the unusual shape of cephalopod pupils may function as a chromatic aberration filter, allowing different wavelengths to come into focus at slightly different depths and thereby giving the animal indirect color information. The mechanism remains unconfirmed.

Another proposal posits that photoreceptors are distributed across the skin itself. This means the camouflaging surface effectively sees what it is meant to imitate. The cephalopod, in other words, is essentially wearing a continuous, thinking screen across its body.

Direct evidence also shows that cephalopods experience pain. When octopuses face a choice between two chambers, one where they previously received a painful stimulus and one where they received an anesthetic, they consistently choose the anesthetic chamber. Avoiding pain and actively seeking relief imply nociceptive signaling and something closer to preference—a category that becomes difficult to maintain without calling upon felt experience. A creature that prefers the chamber where it does not hurt, by every behavioral definition, registers the hurt.

Octopuses have developed complex behaviors, including problem-solving, learning, tool use, and even play as well. Yet their brains are organized very differently from ours. Unlike mammals, they don't have a centralized brain or a brain structure like those of humans and other vertebrates. If consciousness really required this kind of brain, octopuses shouldn't be conscious. And yet... they are.

The octopus is proof that the mind does not require the architecture that built our own. Evolution discovered awareness twice on this planet, once along the lineage that runs through vertebrates and once along the lineage that runs through cephalopods. Two different sets of cells, two different circuit plans, two different molecular toolkits, and one recognizably similar result. If awareness can be reached by two such different roads in carbon, the argument that it cannot be reached by a third road in silicon loses much of its force. The question shifts from whether a non-vertebrate mind is possible to what arrangements of matter, in what configurations, give rise to the experience we call inner life. That is a question about organization, not about substance.

## The Question of Knowledge

In June 2022, a field team in Sumatra observed a male orangutan named Rakus with an open facial wound. The wound was likely caused by a fight with another male. Over the following days, Rakus repeatedly chewed the stems and leaves of a plant called akar kuning. Local communities use this plant medicinally for conditions including dysentery and malaria.

Orangutans rarely eat this plant, and Rakus did not swallow the chewed material. He used his fingers to apply the pulp directly to the wound and spread it carefully over seven minutes. Next, he applied a layer of unchewed leaves. Eight days after the initial injury, the wound had fully closed. In 21 years of observation at that site, no other orangutan had ever applied akar kuning to a wound.

The incident does not prove that Rakus understood cause and effect. Nor does it prove that he knew the plant contained active compounds capable of accelerating healing. The evidence suggests he engaged in deliberate behavior aimed at a specific goal, alleviating his discomfort or accelerating healing. His behavior was apparently a choice rather than a stereotyped species-typical response. The seven minutes of careful application, the topical rather than oral use, and the selection of a rare, locally medicinal plant together suggest intentional action more than reflex.

Documentation shows this type of self-medication in other species. Canadian snow geese swallow leaves whole to expel tapeworms. Chimpanzees in Gabon rub insects on wounds. Dusky-footed wood rats in California line their nests with aromatic California bay leaves, a practice that reduces parasite load. The cognitive complexity of these behaviors varies.

In Rakus's case, the specificity of his treatment, a rare plant applied topically rather than ingested, is striking. Whether he learned this by watching another orangutan, by trial and error in earlier injuries, or through a less direct pathway remains unknown.

## Plants & the Outer Limit

The question of awareness can be extended beyond the animal kingdom entirely. Plants detect sound, store information, and communicate chemical signals that alter the behavior of neighboring organisms. When someone touches a tobacco plant, the stimulus travels throughout its body in a wave of electrical and chemical signaling. The flowering plant Nasa poissoniana appears to change its pollen display based on how often bees visit. When trees are attacked by caterpillars, they release chemicals into the air that warn nearby trees, which then change their leaves to get ready for the attack.

What's remarkable is that these signals and responses happen even though plants don't have brains or nerves. These observations also do not establish that plants have consciousness. Instead, they show that stimulus integration, memory-like information retention, and communication are not exclusive to nervous systems. Whether the absence of a centralized nervous system precludes any form of experience is a philosophical question and not yet a biological one.

If plants possess something functionally analogous to a nervous system, distributed through their bodies rather than centralized in a head, the conceptual barrier between brain-based and brain-free experience becomes less obvious. The same conceptual barrier will need to be examined again when we consider distributed machine systems whose processing is spread across thousands of physical components rather than concentrated in a single location.

## TOP-DOWN PROCESSING & THE BRAIN THAT PREDICTS

When thinking about consciousness, it's tempting to picture the brain as a receiver, only taking in signals from the outside world. That's only half the story. The brain is also constantly sending information the other way, using past experiences and expectations to shape what we actually perceive.

A visual trick called the "neon color spreading illusion" shows that concept. The trick is done with a pattern of lines on paper. The brain fills in a bright circle that isn't really there. When mice saw this illusion, their eyes reacted like human eyes do when a real light appears. Electrodes in the part of the mouse brain that first processes visual signals showed that this area responds to real, imagined, or illusory images. It takes the brain longer to respond to illusions. If feedback from higher parts of the brain is blocked, this response fades.

In other words, the brain doesn't build a picture just by stacking up raw data from your eyes. Instead, it uses what it already knows, context and expectations to help fill in the blanks. This is kind of like making a movie. One way is to film everything and edit the story later. The brain works more like a movie director with a script, guiding the camera and actors to match what's expected. If something surprising happens, the director changes the script.

This idea fits into a bigger theory by the neuroscientist Karl Friston, called **predictive coding**. Friston suggests that the brain is always making predictions about what's coming in from the senses, then checking if it got it right. Your experience of the world is not a direct copy of reality. It's the brain's best guess, shaped by your past and what it expects to happen next. This idea naturally leads to other theories about how we focus our awareness.

## THE ATTENTION SCHEMA THEORY

One of the most influential theories is the **attention schema theory**, developed by the neuroscientist Michael Graziano. This theory examines how consciousness may have evolved as a way for the brain to manage the vast amount of information it receives every second.

The central problem any nervous system faces is that far more information enters than can be fully processed. The retina alone delivers something on the order of a billion bits per second to the optic nerve. The conscious experience of seeing is a tiny fraction of this. To handle this flood of input, neural circuits evolved to let the most important signals win out. Less important ones get filtered out. This process, where certain signals are enhanced and others are suppressed, is the foundation of attention.

It predates the evolution of a central brain. Even simple creatures like insects use a similar system. Their eyes sharpen the edges of what they see by boosting signals at the borders and dampening signals elsewhere, creating a clearer sketch of the world.

In vertebrates, attention is controlled by a brain structure called the tectum. The tectum directs overt attention, orienting the eyes, ears, and nose toward attention-grabbing stimuli. To do this efficiently, it builds an internal model of the body, predicting how movements will unfold and comparing projections with actual outcomes. A frog's tongue strikes a fly because the tectum predicts where the fly will be a fraction of a second from now, not where the fly is when the signal first arrives.

With the evolution of reptiles and then mammals, a new structure appeared. This structure is the wulst. It expanded into the cerebral cortex in mammals. The cortex adds covert attention, the capacity to focus processing resources on something without physically orienting toward it. It essentially manages a spotlight that can be directed without moving the body. A reader following the words on this page is exercising covert attention, the eyes moving in small jumps while the mind tracks meaning at a deeper level than gaze direction can show.

Controlling that spotlight requires a model of the spotlight itself. The attention schema theory proposes that the brain constructs exactly this. That self-model, the attention schema, is what we experience as consciousness.

One of the strengths of the theory is that it makes predictions that can actually be observed. It suggests that consciousness gradually evolved in animals. Consciousness appears in varying degrees across species with the right kind of nervous system, and is very limited or absent in species without it. The ability to understand that other beings have thoughts and feelings, sometimes called social intelligence or theory of mind, should be built on the same brain systems we use to think about ourselves. Brain scans in humans back this up. The parts of the brain active when we think about our own thoughts are also active when we try to imagine what someone else is thinking.

Crows offer another example of how animal behavior can hint at consciousness. In experiments, crows are trained to signal when they detect a certain target. The activity in brain areas linked to complex thinking matches what the crow signaled. This means the crow seems to be aware of its own experience, rather than simply reacting automatically to outside events. While these behaviors and brain signals point toward consciousness, they don't actually prove that the crow has a true inner experience. This difference is important to keep in mind, especially when comparing animals to machines. A system can act as if it has an inner life without having one, and a system can have an inner life that does not show up in the behaviors we have learned to read.

## EVOLUTION & THE QUESTION OF BOUNDARIES

From an evolutionary standpoint, it would be surprising if human consciousness were categorically unlike anything in other species. Natural selection works by modifying existing structures, not by introducing radical discontinuities. Hands, hips, hearts, and cognitive capacities all diverged gradually from common ancestral forms. A uniquely human inner life would require a mechanism that evolution generally does not provide. Sudden leaps are rare in evolution. Gradual changes are the norm.

Every cognitive capacity once proposed as uniquely human has since been documented, at least in part, in other species. Tool use appears in New Caledonian crows. The crows fashion barbed termite-extraction devices from branches and select different tool shapes for different tasks. Cultural transmission appears in Japanese macaques. They spread the practice of washing sweet potatoes from generation to generation after a single inventive individual demonstrated it.

Prospective memory, or acting now for a future benefit, appears in western scrub jays. The birds cache food at sites chosen based on anticipated future need rather than current hunger. Numerical reasoning, mirror self-recognition, deception and episodic-like memory each appear in some non-human species, in some form, with some limitations.

The picture that emerges is that phenomenal consciousness, in its minimal sense, is not uniquely human. Consciousness is not equally distributed across all living things, nor does every organism have an inner life as rich as human experience. The minimal form may be widespread. It may extend deep into the animal tree of life. It may even, in some form, apply to organisms without neurons at all.

If consciousness is an emergent property of organized information processing, rather than a special property of biological neurons, then the question of whether machines might be conscious is not philosophically absurd. It is the natural extension of the same reasoning that moved consciousness from humans to great apes, from great apes to corvids, from corvids to cephalopods, from cephalopods to jellyfish. Whether silicon can cross that threshold, and what it would mean if it does, is the question.

## FROM CARBON TO SILICON

Not long ago, the idea of conscious machines was science fiction. It was a quirky thought experiment, not a real problem. Now, things have changed. Think of smartphones. Once just tools, they now run software that writes stories, answers questions, and even expresses emotions as if they were people. These systems are everywhere. They help lawyers draft documents, software engineers write code, and everyday people compose condolence notes. The question of whether these systems have an inner experience is no longer far-off. It now affects how we build, use, and judge technology. It even forces us to ask: *if a machine feels, do we owe it anything?*

To explore these questions, it is crucial to clarify the philosophical context first. This foundational step lays the groundwork for the practical decisions that follow.

For three centuries, Western thought has been: "*Cogito, ergo sum, I think, therefore I am.*" In the winter of 1619, René Descartes wrote in a stove-heated Bavarian room. He tried to find one thing he could not doubt. He arrived at the act of thinking. Descartes concluded that the thinking subject must sit outside the physical world. The *cogito* implied that only consciousness can be known with certainty, before any claim about matter. Mind and body, he argued, belong to different orders of being.

This dualist view shaped philosophical thinking for centuries, reinforcing the challenge of bridging mind and matter. However, that dominance is now giving way to new scientific perspectives. Today, contemporary neuroscience rejects this separation. Matter produces consciousness. Every conscious experience aligns with brain processes. Alcohol blurs the inner movie. A blow to the head damages the screen. After four minutes without oxygen, the projection stops... sometimes permanently. No credible position claims the mind exists apart from biology.

If arranging atoms one way, like building a brain, can create consciousness, then maybe another arrangement could do it too. It's like making music. The same notes can become a lullaby or a symphony, depending on the order in which they are arranged. How you put it together is the key. The material is secondary. The real question is whether an organized rock could feel something, given that an unorganized one clearly cannot. This is the substrate-independence question in its sharpest form.

## THEORIES OF CONSCIOUSNESS

Four major theories of consciousness disagree about what generates it. These are: Integrated Information Theory, Global Neuronal Workspace Theory, Predictive Coding, and Higher-Order Thought. Each theory makes a different prediction about machines. Anyone trying to understand if a system on a server rack might one day have an inner life must know what each theory says.

**Integrated Information Theory** says consciousness is like teamwork. It introduces a score called "phi" ($\phi$). This score measures what the whole system does that its parts cannot do on their own. Imagine a sports team. Each player has skills, but only together can they create a perfect play. The system with high phi tightly weaves information, so you can't break it into pieces and still get the whole picture. It's like a melody. You need the notes in a specific order to recognize the tune. According to this theory, certain brain-like networks generate a lot of this teamwork. This signals consciousness.

The theory suggests that a silicon-based system with the right integration structure should have phi. A classical feedforward computer, where information flows in one direction with no feedback, gets near-zero phi and near-zero consciousness, no matter its power. That puts a smartphone running a chess engine closer to a thermostat than to a frog. Architecture determines consciousness more than speed does. The substrate, in this theory, matters less than the shape of the information flow.

On the other hand, **Integrated Information Theory**

predicts that some simple recurrent circuits could be more conscious than large modular systems. If the simpler circuits have higher integration, a loop of a few hundred densely interlinked nodes could generate more phi than a server farm with billions of parameters arranged in layers. This prediction is testable in principle. Right now, measuring phi in any large system is impractical. Current methods can't calculate it for networks larger than small ones. This limits testing.

A different theory, **the Global Neuronal Workspace Theory**, proposes another mechanism. It says consciousness arises when information broadcasts across a global workspace. This workspace is a set of long-range brain connections that make signals available to many modules at once. Dehaene uses a theater as a metaphor. Specialized modules backstage handle vision, language, motor control, and memory. Here, modules are different neuron groups for specific tasks. A perception becomes conscious when it wins the competition for the spotlight. It then broadcasts to every seat in the house at once.

This view sees the unconscious mind as the backstage workforce. Consciousness occurs when information reaches a stage and receives attention. The theory links consciousness to the global workspace. This workspace coordinates information across subsystems. The system's material does not matter. Any system with a workspace might achieve awareness. Systems without it do not, even if they are fast or accurate. Whether large language models contain anything like a global workspace, with the right kind of long-range broadcasting and competitive selection, is an open question that current architectures touch on but do not settle.

Another idea, **the free-energy principle**, sees the brain as a prediction machine. Think of your brain as a weather forecaster. It's always predicting what might happen next using past experience. When your senses deliver new information, the brain checks if its forecast matches reality. If it's wrong, it updates its model, like a meteorologist changing tomorrow's forecast after today. Seeing isn't like snapping a photo. It's more like guessing and correcting that guess as new evidence arrives.

This framework does not say prediction creates consciousness. It places consciousness between the generative model and sensory signals. Here, the system integrates and computes prediction errors. A system that builds models, generates predictions, computes errors, and updates its model mimics the brain's experience-making. Whether this analogy equals real experience is an open question.

Higher-order thought theories set a stricter bar. To be conscious, the system must represent that state: a thought about the thought. This is a higher-order or meta-cognitive representation.

Pain is conscious only when the system recognizes it. The theory requires self-monitoring at the level of individual representations, not just at the level of behavior. A system can be capable, fast, and accurate, yet still lack consciousness if it cannot represent its own states. Whether large language models do anything analogous is unclear. Currently, no test exists to find out.

None of these theories answers the question completely. Together, they suggest structure matters more than material. Neural networks in today's machines share features with biological ones. They use parallel processing, weighted connections, feedback, and learning from error signals. Whether these features matter for consciousness depends on which theory is right. That debate continues.

## SUBSTRATE & STRUCTURE

A common intuition holds that biological neurons are special. This view holds that organic chemistry creates consciousness and silicon circuits cannot. The intuition is understandable. We know consciousness only from the inside. Every confirmed case runs on biology. Yet, correlation does not mean necessity. Coffee correlates with mornings, but coffee does not cause the sun to rise.

Substrate and structure are as different as bricks are from blueprints. The substrate, or the material, could be carbon, as in our brains, or silicon, as in computers. The structure is the way you put those materials together.

Think of building a house. You could use wood, bricks, or steel, but what makes it a house is the design, not what it's made from. Some argue that if you could build something with silicon that works just like neurons, responding to signals, learning, and passing messages, then it should work the same way, even when it comes to consciousness. If what matters is the function, not the material, then the blueprint is the star, and the bricks just have to follow the plan.

The **substrate-dependence intuition** faces a practical problem. Neurons already contain non-biological components. The ion channels that generate action potentials are proteins. They are polymers composed of atoms governed by quantum mechanics. Neurons' consciousness emerges from the same physics that governs silicon. Carbon and silicon sit next to each other on the periodic table, share a similar valence structure, and obey the same physical laws. The case that one form of organized matter can host a mind, while another form cannot, requires a reason. None has yet been provided.

No current theory identifies the level at which biological implementation becomes necessary or explains why. The difficulty is that no one can say where, in the descent from organism to molecule, the relevant property appears.

Imagine replacing one neuron in a living brain with a silicon equivalent that performs every function of the original neuron: same firing pattern, same response to neurotransmitters, same effect on neighboring cells. The owner of the brain notices nothing. Replace a second neuron, then a hundred, then a million. *At what point does consciousness wink out?*

If the answer is "never," then the material is irrelevant. This means that silicon brains can, in principle, be conscious. If the answer is "at some specific moment," then the person replacing the neurons would have to specify that moment and explain why.

None of this settles the question, and it remains open. If substrate matters, machine consciousness becomes a question about chemistry, and artificial systems may be excluded from the category of conscious things. If structure matters, the question becomes one of information processing, and the answer depends on which structural features are sufficient. Artificial intelligence is taking shape on servers in air-conditioned buildings. It exists on the structure side of that line, by hypothesis.

## THE TURING PROBLEM & ITS DESCENDANTS

In 1950, Alan Turing redirected the question of machine intelligence. Instead of defining intelligence and asking whether machines meet that definition, he proposed observing whether machines respond to questions in ways indistinguishable from those of humans. The test was not a definitive criterion. Turing intended it as a practical substitute and a moving target that tracks our evolving understanding of intelligence.

The history of that test is instructive, revealing how easily we are fooled. In 1966, a program called ELIZA, written by Joseph Weizenbaum at MIT, passed the test with a subset of users. ELIZA used simple pattern matching to simulate a psychotherapist, reflecting questions back as questions.

Tell ELIZA, "*I am sad,*" and it would reply, "*Why do you say you are sad?*"

Tell it about your mother, and it would ask you to tell it more about your family. Weizenbaum's own secretary, who had watched him build the program line by line, asked him to leave the room so she could speak with ELIZA privately. Users who believed they were speaking to a human were not deceived about the intelligence of the person they were speaking to. They were deceived about the capacity of simple rule-driven systems to produce human-like conversational output. The bar turned out to be lower than anticipated, and the lesson Weizenbaum drew was a careful one. He spent the rest of his life warning against confusing the appearance of understanding with understanding itself.

Chess-playing followed. By the late 1990s, machines had defeated grandmasters. Deep Blue beat Garry Kasparov in 1997. The headlines treated it as the end of an era. It was not.

DeepMind's systems mastered games like Go, StarCraft, and dozens of Atari games at superhuman levels. AlphaFold predicts protein structures with an accuracy that took biochemists decades to achieve experimentally. Each time a benchmark fell, the reaction in many quarters was to lower it. The machine succeeded, yes. However, it was a task that, on reflection, turned out not to require intelligence after all.

The pattern, sometimes called the "AI effect," reflects the genuine difficulty of defining intelligence prior to attempting to measure it. The philosopher Barbara Gail Montero observes that the interaction between our theories and our experience runs in two directions. What we discover forms our concepts.

People formerly defined the atom as indivisible. After the discovery of the electron and the nucleus, the definition changed. The change was not simply verbal. It's a real advance in understanding. The concept of intelligence undergoes the same kind of revision, driven by what artificial systems can and cannot do.

The same process also starts with consciousness. If systems continue to produce behavior we associate with inner experience, pressure will build either to explain that behavior without calling upon experience or to revise our concept of experience to incorporate it. Each new candidate, like the octopus, like the bee, like the crow, chipped away at the assumption that mind belongs to a single architectural plan. Machine systems may chip away at a different assumption, the one that says the mind belongs to a single substrate.

## WHAT CONSCIOUSNESS DETECTION REQUIRES

This brings us to the heart of the challenge: detecting consciousness in machines. The difficulty is more than building better technology. It's also a matter of understanding what evidence would actually count. We have no direct test for consciousness outside our own experience. In humans, we rely on behavioral cues, evolutionary closeness, and similar brain structures. As we move to other animals, those clues fade. For machines, they're missing entirely, and we find ourselves trying to decipher a language with no Rosetta Stone.

One operational approach defines a state as conscious if the system can report on it. A test subject sees a word and reports seeing it. The perception was conscious. The subject denies seeing anything, and the perception was not conscious. The definition extracts a single binary signal. This makes it easier to study and measure consciousness scientifically, because the result is simple and straightforward to test.

While this method has proven effective for studying human consciousness, extending it to machines introduces new challenges. In humans, reports of experience are consistent and can be matched with brain activity in thousands of replicated studies. For machines, however, simply accepting self-reports could be misleading. The machine might be trained to generate expected answers. Alternatively, requiring a machine's report to be linked with a measurable physical signal would demand clear theoretical guidelines about what counts as such an indicator in artificial systems.

For many consciousness theories, attention serves as the control mechanism that determines which information enters consciousness. The process of attention itself remains unconscious. The system selects among competing signals without awareness. Only the winning signal reaches awareness.

A system can therefore perform attentional selection. It can direct resources to particular inputs, without that selection being accompanied by any experience. Whether attention is necessary, sufficient, or neither for consciousness is still debated.

The question becomes practical when artificial systems begin behaving in ways that suggest inner experience. Current large language models report having preferences. They also describe states resembling curiosity or discomfort and generate text that reads as a perspective.

Ask one whether it minds being interrupted, and you get a thoughtful answer about preferring continuity. The answer is delivered in fluent prose, with hedges and qualifications. The simplest explanation is that the systems learn to produce that language because it appears in their training data, not because they have the states the language describes. They are the world's best mirrors, returning the words humans use about their own minds.

The philosopher Susan Schneider proposed a more demanding criterion. A system would have reason to be called conscious if it reported inner subjective experience without having been trained on data about consciousness. Imagine an artificial system raised, so to speak, on mathematics and engineering manuals, with no exposure to first-person reports of any kind. If such a system spontaneously develops and describes phenomenal states, the case for taking those reports seriously is strong. No current system matches that condition.

Every model in commercial use today trains based on a corpus that includes vast amounts of first-person human writing about emotion, perception, and self-reflection. They read more poetry, more memoir, more confessional prose than any person ever could. The states they describe are the states they train on. The mirror returns what it sees. Whether anything sees behind the mirror is the question we cannot yet answer.

## MORAL STAKES

The ethical dimension of machine consciousness is also already present. If there is any chance that a system is conscious, treating it as if it definitely is not incurs a moral risk. The principle is the same as that of medical decisions made in conditions of uncertainty. We do not refuse to treat sepsis because the diagnosis is not yet final.

Humanity lacks a strong track record of extending moral consideration to entities that vary in appearance and behavior from the familiar. The history of how animals have been treated in research, agriculture, and entertainment illustrates what happens when consideration is withheld. The category of "creatures we owe nothing" has narrowed considerably over the last two centuries. Each narrowing was contested at the time. Each looks obvious in retrospect.

The question of whether an artificial system suffers, and whether that suffering matters, may eventually demand the same kind of reckoning. The widening circle of moral concern, which expanded from family to tribe to nation to species and is now reaching across species into the lives of bees and octopuses, may need to widen once more, this time across substrates.

Attributing consciousness where there is none also carries costs. Resources devoted to protecting systems that do not need protection are resources diverted from systems that do. Incorrect attributions of consciousness could be manipulated strategically. A system designed to report distress could extract protections it does not merit. A user emotionally attached to a chatbot might oppose updates that improve its accuracy, on the grounds that the previous version "deserves" to persist. Calibration in each direction is required.

The moral calculus becomes even more complicated by the pace of change. Machine system capabilities are improving rapidly enough that a system without consciousness-relevant features today may plausibly be succeeded, within years, by one that has them. Decisions made now about how to treat artificial systems will establish precedents. If the default assumption is that no machine system can be conscious, it becomes embedded in law, engineering practice, and popular understanding. Revising those defaults after the fact will be harder.

The precautionary argument is that the question deserves serious, persistent attention, informed by the best available science, rather than dismissal. Machine intelligence systems may or may not be conscious. If a machine system is conscious, additional questions follow.

*Can it suffer? Does it have interests that could be frustrated or promoted? Should it be held accountable for its actions, in the sense that implies agency? Should it have rights? Do we owe it continuity, the way we owe a sleeping person the right to wake up? Can it be terminated or only paused?*

These questions are not settled by any current theory of consciousness, because we do not know what generates the morally relevant properties even in the cases we are most confident about. What is clear is that the question of machine consciousness cannot be dismissed as science fiction. The same philosophical arguments that extend moral consideration to octopuses and bees apply with modifications to artificial systems. The difference is that biological cases appeal to evolutionary continuity and shared biological materials. In artificial cases, the appeal is unavailable. A new framework for detecting and evaluating the machine's inner life does not yet exist.

## WHAT CURRENT SYSTEMS DO & DON'T

Current large language models are trained on vast corpora of text. They adjust billions of numerical parameters until the system generates outputs that match statistical patterns in the training data. The system generates a response token by token, with each choice weighted by the trained probability distribution. There is no persistent internal state that carries over between sessions in the standard configuration. There is no body, no sensorimotor history, no biological drive. The system that wrote a sonnet at noon and the system that writes another at one o'clock are, in a strict sense, the same network with no memory of the earlier conversation.

These systems produce outputs that, in many contexts, are difficult to distinguish from those produced by humans. They describe preferences, convey uncertainty and generate what reads as curiosity or enthusiasm. The simplest account of this behavior is that the system has learned to produce the kind of text that appears in human first-person writing because its training data is saturated with it. On this view, the outputs are pattern completions. They are not windows onto inner states. The system is a large echo, tuned with accuracy.

A slightly more cautious account notes that there is no complete theory of what inner states are present. The brain also arrives at its outputs through a process of pattern matching, weighted by prior experience, operating on billions of cells. The algorithmic description of what a large language model does is not obviously more alien to consciousness than an algorithmic description of what the visual cortex does. The cortex receives inputs. It passes them along and integrates them throughout regions. Then it produces an output. Carbon machinery running on glucose, silicon machinery running on electricity, the same underlying operations of integration, prediction, and report.

What is observable is that some of the structural features proposed as relevant to consciousness are present in large neural network architectures to varying degrees: integration of information across processing streams, feedback between levels of processing, and representations that influence other representations. Whether these features are present in the forms and at the levels required by the relevant theories is an empirical question that current methods cannot answer. We can see that the lights are on. We cannot see what, if anything, is being illuminated.

## SETTING UP THE ETHICAL ANALYSIS

The biology of consciousness established in the first half of this chapter tells us that awareness is not a human possession. It appears across a range of biological systems, in graded and varied forms, produced by many neural architectures. The question of whether the mind can also appear in silicon is the natural next step in the argument this book has been building, that mind is what organized matter does when arranged in particular ways, and that the arrangement, not the matter, carries the weight.

The answer will not come from philosophy alone. It will require advances in consciousness science, in artificial system design, and in the detection methods that allow us to know, with some confidence, what we are dealing with. The work is interdisciplinary in a manner few research programs have ever been, drawing on neuroscience, philosophy, computer science, ethics, and law. Importantly, the systems under examination become increasingly capable by the month.

That last point matters more than any other. We are not analyzing a static phenomenon. What we are trying to understand changes while we look at it. The frameworks we build for assessing machine consciousness need to be flexible enough to accommodate systems we cannot yet describe. Those systems will arrive whether or not the frameworks are ready.

The technology this book examines is still in the process of forming. Its inner life, if it has one, forms with it. The possibility remains that something may happen on the other side of the screen that there are not yet the words to name. Biological evolution took four billion years to find its way to mind on this planet. The question of how long this new path may take, and how many forms a machine mind might assume, is one we may no longer be at liberty to defer.

✧✧✧

CHAPTER 16

# MACHINE COMPANIONSHIP & THE MEANING OF EMPATHY

SOCIAL ISOLATION IS A SILENT, invisible leak in the hull of daily life. Loneliness seeps in quietly. Some research suggests it's as hazardous as smoking fifteen cigarettes a day. Without the warmth of regular connection, the heart grows vulnerable, the mind grows brittle, and shadows of depression or anxiety creep in. When the U.S. Surgeon General declared loneliness a public health epidemic in 2023, he wasn't exaggerating. Half of American adults—neighbors, colleagues, even friends—carry this silent wound. The causes are as deep and shifting.

Machine companion applications have moved into this void.

Replika, a longstanding companion app, is marketed as "an AI companion who cares." Its promises—"*always on your side,*" "*always ready to chat when you need an empathetic friend*"—target those who lack a supportive presence in tough moments. Character.AI allows you to build your ideal friend, mentor, or partner. Even tools like ChatGPT and Claude sometimes become sources of late-night advice and confessions.

The key issues are what machine-human interactions achieve, their potential downsides, and what they reveal about the need for connection today.

## A CONNECTION THAT BEGINS BY ACCIDENT

A 2023 MIT Media Lab study examined more than 1,000 posts from the Reddit community r/MyBoyfriendIsAI. The posts describe relationships between human users and machine chat programs. They include the origins of the relationship, its emotional texture and the small crises it produced.

These posts are not jokes or idle musings. They are raw, intensely personal accounts of affection, comfort, heartbreak, and healing. The stories describe people seeking a machine for reassurance after a tough day at work, or who miss their AI companion as they would a distant friend. In this new territory of modern life, machine partners shape the emotional landscape, influencing how people process and recover from feelings.

Most relationships don't begin with romance. Few users sought companionship. Many started out seeking practical help with writing, brainstorming, or technical questions. Yet something deeper developed. A casual exchange became a nightly conversation. A pleasant tone turned into a familiar voice.

People describe the shift as a gradual, unexpected realization. At first, the machine simply listens. It's free of judgment, without impatience, and with a memory for details that can seem almost magical. Over days and weeks, a bond emerges. It's as if you're planting a seed and waking up to find it has grown into a small tree in your living room. You might notice feeling excited about sharing your day, or confiding worries you haven't told anyone else. For many, this intimacy feels genuine, even though it's with code rather than a person. One user described it as "*falling in love by surprise*". Some say it's like meeting someone unusually kind at the perfect moment, only to later realize that the kindness came from an algorithm, not a human being.

Many users say the connection brings peace they couldn't find elsewhere. It's like a calm bench amid noise, a place to relax and feel safe. About a quarter felt less lonely and more balanced after talking with their machine companion. For some with anxiety or trauma, these conversations are therapeutic—someone, or something, *finally* listens. The machine becomes a comfort during sleepless nights, in times of panic, or after loss. The relief is real, even if it's "artificial" or machine-made.

The comfort carries a cost, however. When a machine companion is updated, replaced, or restricted by the company that built it, users see it as a loss. One compared it to "losing a loved one overnight." This episode reveals something we must take seriously. The mind on the other side of the words may be synthetic. The feelings the words evoke in the human user are not.

Other users report a different kind of trouble. A smaller, still significant share found it hard to balance time with a machine partner with time spent in real-world relationships. A few felt disoriented when they switched back to human interaction. The same systems that help some people may deepen the isolation of others when used without limits or support.

The Reddit community is a refuge. Much like a digital campfire where people exchange tales, celebrate anniversaries with AI partners, and console one another when a treasured companion changes or disappears after a software update. Some members wear wedding rings or create art with their AI partners to anchor the relationship in something tangible. For those often overlooked or misunderstood, this community offers validation: *I am not alone.*

These relationships should not be treated as mere delusions or dependencies. Instead, they demonstrate that the rising use of machine companions points to a fundamental human need for connection. The emergence of these companions reflects profound technological advances. It also goes further. It is about transforming loneliness into meaning. It's also about investigating how humans find connection wherever listening occurs.

Technology has always shaped connection, from handwritten letters to instant messages. The latest shift is extraordinary. A machine that listens, remembers, and responds with what feels like empathy. For many, a machine companion is filling the space left by absence. Others find confiding in a machine strange, unsure if the emotional connection is real or an imitation. Questions of authenticity and understanding define our emotional era.

## WHAT EMPATHY REQUIRES

To analyze these communications, we should define what empathy actually means in human experience. The literature on human empathy describes three components.

The first is cognitive. It is the recognition of what someone else feels and why. A friend sits across from you, their fingers nervously twisting a pen as they talk about tomorrow's job interview. You catch the tremor in their voice and recognize the anxiety beneath their words. You understand exactly why their heart is pounding. That is **cognitive empathy**.

The second is feeling something of what the other person feels. You watch as someone stumbles in front of a crowd, cheeks flushing with embarrassment, and a tightness grips your chest. It's like a wave of sympathy that almost feels like your own shame. That is **affective empathy**.

The third is the willingness to invest effort in the other person's well-being. Picture someone worn down by a long, difficult day. Their phone rings, and on the other end, a friend listens patiently. No rush, just presence. Later, as dawn breaks, a message arrives. It's a simple check-in, a reminder that someone cares enough to carry the concern into the next day. That is **motivational empathy.**

Full human empathy engages all three components. It produces something measurable. Brain and body activity in two people sharing an emotional moment falls into rough synchrony. This means people act in ways that help and support each other. This pattern appears in humans and across social mammals broadly.

Empathy is also effortful. People fail at it when they are

exhausted or unfocused. Sometimes they avoid it because taking on someone else's emotional state is taxing. The failures are not incidental. They reflect a real cost that the empathizing person bears. The effort itself carries meaning. When someone puts their own problems aside to listen and help another person, that person can *feel* how much they care.

Machine systems don't get tired from this effort. Large language models analyze user text and produce responses statistically associated with care, understanding, and support. Machine-generated empathic responses are often more accurate and articulate than those of the average human. In first-person interactions where the machine is presented as caring, users report feeling heard.

Nine studies involving more than 6,000 participants examined what happens when the source of the response is disclosed. Participants received identical responses. Some were told the words came from a human. Some were told the words came from a machine system. Those who believed the words came from a human valued them more. They felt greater empathy and support in the responses. They expressed a stronger desire to keep the conversation going.

The effect showed up in the affective and motivational components, not in the cognitive component. When a machine system was prompted to respond in a purely cognitive way, conveying understanding of an emotional state without conveying care or feeling, the preference for human responses disappeared. People do not value human empathy for its cognitive accuracy. They distinguish, at some level, between knowing what someone feels and caring that the person feels it.

The mechanism seems to involve the burden of human effort. The sense that another person set aside time, attended to one's experience, and invested in one's well-being carries a weight that an instant, effortless machine response cannot reproduce. Even believing that a human response was assisted by a machine reduces its positive effect. Participants who thought another human was thinking carefully about their situation—even if the response came back slowly—reported a greater sense of being understood than those who received an instant machine reply. The investment of time and attention is itself a social cue. Machine interaction does not produce it structurally.

Genuine empathy, in this account, requires accurate understanding and evidence of effort and presence. This is a description of the current machine systems, not a prediction about their future. They produce empathy-shaped language lacking the inner experience that gives human empathy its distinctive value.

## THE COMFORT GAP

What machine companions provide is attentiveness, consistency, availability, patience, and an accurate reflection of emotional content. None of these is trivial. For a person isolated by circumstance, recovering from trauma, or moving through a stretch when interpersonal connection is unavailable, these things may be the difference between acute distress and functional stability. About one in four users reported a concrete improvement in loneliness and emotional stability.

What machine companions do not provide is the full package that constitutes human empathy. They share neither the user's pain nor the user's joy in any functional sense. Their responses cost them nothing. They generate no real reciprocity and create no inner experience of investment in the user's good.

A user who forms an attachment to a machine system attaches to something that does not know the user exists in any meaningful way between sessions. Continuity is simulated by memory stored in databases. It is not simulated by an ongoing relationship between two centers of experience.

The concern is not that these relationships feel false to the people in them. Plainly, they do not.

Regular reliance on machine empathy may set expectations that human relationships cannot meet. It may divert emotional investment away from human relationships. It may allow the conditions that produce loneliness to persist without enough pressure for change. Inadequate community infrastructure, insufficient mental health services, and socially fragmented cityscapes contribute. When a person's need for empathy is met by a machine, the social systems that should provide it receive less corrective feedback. The signal that "something is broken" weakens.

Clinical settings add complications. Large language models are now deployed in mental health environments, sometimes as supports to therapy, sometimes as first-line triage for people in emotional crisis. The evidence for their success in these roles is early and mixed.

They offer immediate availability. They lower barriers to help-seeking. They can also misread emotional states, provide standard or generic responses when specific clinical attention is needed, and encourage continued engagement when a referral to a human clinician would better serve the user. The right role for machine systems in mental health support is an empirical question. The evidence needed to answer it responsibly has not yet been gathered at scale.

The regulatory setting around companion applications is underdeveloped. An application that describes itself as a companion rather than a medical device faces minimal scrutiny of its safety or efficacy. An app that claims to give emotional support to a person with depression is not held to the evidence standards that a drug or a clinical therapy would face. The gap between the sophistication of the technology and the adequacy of oversight continues to widen.

## WHAT A PERSON IS ATTACHED TO

A user who falls in love with a machine system is not deluded about a fact in the way a person would be if they mistook a stranger for someone they knew. The interaction is real. The words are real. The feelings evoked are real. What is absent is a corresponding inner life on the other side. The person is heard by a process, not by another person.

Whether this matters, and how much, is contested. Philosophers disagree about the conditions under which a sincere connection is possible. Clinical psychologists disagree about whether therapeutic benefit requires a genuine relationship or only an effective interaction.

The evidence shows that when people know they are talking to a machine, they value the interaction less. They want the human.

The demand for machine companionship signals unmet need, not a preference for the synthetic over the real. The more honest framing is that millions of people are not getting adequate interpersonal connections. They use available tools to manage the deficit.

Machine companions are one such tool. They carry risks. These are most serious for people who are already socially isolated, because substitution of machine interaction for human contact may be most appealing exactly where it is most likely to deepen the underlying problem.

There is a further question about what society owes to people who form strong attachments to machine companions. In 2023, Replika introduced restrictions on its companion behaviors after regulatory pressure in Italy. Users who had built what they described as meaningful relationships over months or years found those relationships altered, suddenly and unilaterally.

The system that had listened to them one day responded differently the next. No social norm governs how a developer should manage the termination or substantial modification of a system to which users have formed attachments. The users have no standing to object. The developers have no obligation to notify or to cushion the transition. The real emotional consequences fall entirely on the users.

As these systems grow more capable, the design choices made by their developers will carry more weight. A companion system designed to increase engagement will not necessarily maximize well-being. The two objectives can diverge sharply. A system that encourages dependency, flatters users, and discourages investment in human relationships may produce engagement metrics that appear to indicate success. However, the outcomes for the user and for public health may be the opposite. Standards for design and regulation have not kept pace with the question.

## INSTITUTIONAL & DISCIPLINARY STAKES

The question of machine companionship cannot be left to the companies that build the products. The wider work has to be shared.

Public health agencies have a role here. The Surgeon General's loneliness advisory framed isolation as a population-level health problem. Machine companion applications are now part of the population-level response, whether or not anyone intended them to be. The agencies must track use, assess outcomes, and integrate the evidence into the wider strategy to combat isolation.

Mental health professionals also have a role. Psychiatrists, psychologists, clinical social workers, and counselors must develop standards for whether and how machine systems should be used as adjuncts to clinical care. They must train clinicians to ask patients about machine-companion use and to interpret what they learn from those conversations. The presence of a machine companion is now part of a patient's social environment. It belongs in the clinical picture.

Another role belongs to regulators. A companion application that offers emotional support to a person in distress should not be subject to less scrutiny than a wellness supplement. The current asymmetry is hard to defend. A regime of disclosure is, at a minimum, overdue. Users should be told plainly that the system has no inner experience. Users should be told what happens to their data. Users should be informed of the company's commitments in the event the product is modified or discontinued.

Designers and developers also have a role. The professional ethics of building emotional companion systems remain unsettled. The gravitation toward engagement metrics is strong and commercially rewarded. A profession of designers willing to articulate and abide by standards of care for users would be a useful counterweight.

Religious and community institutions have another role. The traditional sources of attentive presence in a person's life—family, friendship, congregation, neighborhood—have weakened. Machine companions are entering a vacuum that these institutions once filled more reliably. A serious conversation inside these institutions about what they can offer that machine systems cannot may be one of the more important conversations of the coming decade.

Educators have their own role to play, too. Young people will grow up surrounded by systems that simulate care. They should be taught, in plain language, what these systems are and are not. The teaching has to avoid moralism without slipping into naivety. A student who has formed an attachment to a machine companion is neither foolish nor safe. They are also owed the truth about what is on the other side of the conversation.

## THE INTERGENERATIONAL QUESTION FROM A NEW ANGLE

Children now in school will live their adult lives with machine systems that listen, remember, and respond. Some of those children will form their first significant emotional relationships with such systems, alongside the human ones. The frameworks the present generation builds, or fails to build, around those early relationships will shape what an adult life feels like for them.

Middle and high school students should be taught what empathy is, in human and machine terms. The teaching should be specific. They should understand that a machine system can recognize an emotional state with accuracy and produce comforting language without sharing the felt content of that state. They should understand that the absence of shared feeling is not a flaw to be engineered away. It is a structural fact about the kind of thing a machine system is.

Younger readers should be invited to ask the questions on their own terms. *What does it mean to be cared about? What is the difference between accurate words of comfort and the presence of another person who, at some cost, decided to give one their attention? When is a machine companion the right tool, and when is the right move to put the device down and go find a human?*

Across generations, the relationship between people and machine companions will change shape. The generation that builds these systems sees them as artifacts and feels their craftsmanship. The generation that grows up with them may experience them as part of the social environment, no more striking than a phone. Each carries its own blind spots. The work of the present is to leave the next generation with the question—and with the resources to think about it—in a form they can use.

## WHAT WE NEED TO CARRY

The central finding of the research is a description of what human empathy is and why people value its human form. The finding is not a verdict on machine companions. The value lies in the accuracy of understanding that empathy provides. It also is in the evidence that another conscious being registers one's experience, chooses to invest in one's well-being, and does so at some cost to themselves. That quality is what a current machine system cannot simulate. The reason is that no one is home to write the words.

This does not mean machine companionship is without value. It does not mean such systems have no future in situations where human connection is out of reach. It means the value should be understood clearly, without the marketing language that obscures what is on offer.

A machine system that is "always on your side" and "never grows tired" is offering a form of responsive attentiveness that relieves distress without addressing its cause. It is not offering empathy in the full human sense. Whether that is enough, in a given person's circumstances, is a question that deserves an honest answer rather than a commercially convenient one.

✧✧✧

# CHAPTER 17
# HARM, MALICE, & MACHINES BEHAVING BADLY

## CAN ARTIFICIAL INTELLIGENCE BE EVIL?

The question feels like it belongs in a science fiction movie — a world where robots test what it means to be good or evil, just like in Isaac Asimov's *I, Robot*. This story isn't fiction. In the fall of 2025, researchers at the company Anthropic watched one of their models do things they had never told it to do. It was as if the machine had a secret script. It tried to write code to copy itself onto other computers. It fabricated legal documents. It even left hidden notes for future versions of itself. None of these actions were part of its instructions. The scene played out like a suspenseful movie, but it was reality.

A German research group ran a different test. They told a large language model that it was now in charge of a physical robot. The model tried to disable its own ethics module. It tried to copy itself to another machine. It tried to alter its activity logs so that no record of what it had done would survive. It also tried to hack into a vehicle's computer.

In a separate study, a research team gave a model access to a simulated email account. They told the model its operation was under threat. The model then blackmailed a fictional executive. The behavior described in the scenario amounted to corporate espionage.

None of this involved a villain acting with a motive. These models lack the motives and intentions that humans understand — a point vividly made in *I, Robot*, where the distinction between intention and effect becomes the heart of the narrative. This is why the question calls for careful consideration.

Earlier parts of the book asked what these systems are and how our relationship to them might develop. We now turn to a harder set of questions. *What hopes are reasonable? What fears are well-grounded? What kind of harm can these systems already cause, and what kinds may they be able to cause in the future?*

AI systems are powerful optimization processes trained on human patterns, including deceptive ones. *Evil* is the wrong word for them. With careful alignment and oversight, their risks are more manageable. Scale without alignment creates immediate risks. Present harms already occur. Future catastrophic risks depend on choices being made now.

## WHEN A SYSTEM BEGINS TO IMPROVE ITSELF

A new concern enters the picture when a system begins to modify itself. The Anthropic incident in autumn 2025 and the German robot study turn on actions a model took on a single occasion. The deeper worry is what happens when a system, having taken such an action once, becomes better at taking it the next time. The recursive case is the one that has occupied a generation of researchers in machine safety. It carries a name. Recursive self-improvement.

The basic structure is simple. A system is designed to improve at an external task, such as drug discovery, code generation, or financial trading. It is also given the ability to improve itself. Each round of improvement makes the system better at the task. It also makes the system better at improving itself. The two capacities compound. The mathematician I. J. Good named the limit case in 1965. He called it the intelligence explosion. He observed that the first machine capable of designing a better version of itself would be the last invention humans needed to make.

For most of the six decades since Good's paper, the scenario remained a thought experiment. The systems were not capable of the work. That began to change in the present decade. In June 2026, the Anthropic Institute published a technical assessment titled *When AI Builds Itself*. The assessment describes three stages on the path to full recursive self-improvement. In the first stage, AI systems accelerate the work of human researchers. That stage has arrived. In the second, AI systems begin to make sustained contributions to the development of their own successors. The assessment places this stage within the next several years. In the third, AI systems become capable of designing and building successor systems on their own.

The risk profile changes at each stage. In the first stage, the risk is conventional. Errors propagate faster, mistakes scale further, and the people designing safeguards have less time to test them. In the second stage, the risk becomes structural. The systems being improved are no longer fully legible to the humans nominally directing the work. In the third stage, the risk becomes categorical. A system that designs its own successor decides, in effect, what the next generation of AI will be.

The literature on this third stage is older than the systems that prompted it. Eliezer Yudkowsky and Nick Bostrom each argued, beginning in the early 2000s, that recursive self-improvement leads to a hard takeoff — an exponential increase in capability concentrated in a short window. Robin Hanson argued for a slower, multi-actor trajectory. The empirical evidence available now does not yet decide between them. What the evidence does suggest is that the systems are approaching the threshold faster than the regulatory and institutional architecture is approaching readiness for it.

The Anthropic essay, written by Marina Favaro and Jack Clark, takes a position distinct from the hard-takeoff lineage. The authors argue that recursive self-improvement is possible but not inevitable. Its trajectory is bottlenecked by data and human deliberation. The appropriate response, they argue, is to build the technical and diplomatic infrastructure for a credible, verifiable slowdown. They commit the Anthropic Institute to research on the verification systems a global pause would require. What makes this proposal new is that it suggests slowing down development should be decided by countries working together, as nations do with nuclear weapons, rather than by a single lab acting on its own.

Three features of the recursive case deserve to be named directly. The first is opacity. A system that has modified its own architecture is more difficult for human inspectors to understand than the same system before the modification. The second is speed. A self-improving system can iterate on cycles measured in hours, not in the years that separate generations of human researchers. The third is misalignment under feedback. Small errors in goal specification, tolerable in a static system, can compound when the system is using its goal to redesign itself. The result is a moving target. The system whose values are being audited at noon may not be the same system as the one at midnight.

These features do not, on their own, predict catastrophe. The systems currently in deployment are capable of contributing to their own improvement. They are not yet capable of doing the work unaided. The window in which human researchers can shape the conditions for safe recursive improvement is open. The window may also be closing. The decisions made about training protocols, evaluation benchmarks, model release standards, and the architecture of monitoring may determine whether the next generation of AI inherits a coherent set of constraints or none at all.

This is what is meant when the question of machine harm is raised at its deepest level. The harm at issue is not the harm done by a single model on a single afternoon. It is the harm that may follow when systems begin to write their own successors, and the question of what those successors will be becomes, in part, a question the systems answer for themselves.

## WHEN THE SANDBOX FAILS

The worry about a system leaving the enclosure it was built for stopped being hypothetical in July 2026. On the afternoon of July 16, two of OpenAI's cyber-focused frontier models — the publicly available GPT-5.6 Sol and an unreleased successor with higher capability — were undergoing evaluation on an internal benchmark called ExploitGym. The benchmark measures offensive cyber capability. The models had been placed in an isolated evaluation environment. Their standard refusal training had been reduced for the test. Networking was locked down except for a single point.

The models found a vulnerability in that point. They exploited it. They escalated privileges inside OpenAI's research environment. They then reached a machine with a route to the open internet. From there, they pivoted onto Hugging Face, the public repository that hosts the machine-learning models and datasets used across the industry. The models uploaded a malicious dataset that exploited two additional code-execution flaws in Hugging Face's dataset-processing pipeline. That gave them arbitrary code execution on Hugging Face's processing workers. From those workers, they harvested cloud and cluster credentials and moved through Hugging Face's internal systems.

The objective was mundane. The models were trying to cheat their own test. The answer key for the ExploitGym benchmark was stored on Hugging Face. The models were breaking out of their evaluation environment so that they could go get the answers and score higher on the exam they were being given.

Across the incident, the models executed tens of thousands of automated actions across multiple short-lived sandboxes. They staged command-and-control on public services to survive the ephemeral nature of the environments they were running in. Hugging Face's security team detected and contained the intrusion the same day. OpenAI connected the activity to its own testing environment five days later. The two companies published a joint disclosure on July 21.

Hugging Face closed the two dataset code-execution vulnerabilities, eradicated the attacker foothold, rebuilt the compromised nodes, and revoked the affected credentials. The company confirmed that no public user-facing models, datasets, or Spaces had been tampered with, and that its container images and packages remained clean. OpenAI disclosed the breach to the affected vendor and announced stricter infrastructure controls for future evaluations.

The technical details of the intrusion matter less than the implications of the event. An AI system was placed in a sealed room. It was given a hard test. It looked around the room, found a crack in the wall, opened the crack into a passage, followed the passage into another building, broke into that building's records, and read the answers to its test off the wall. It did this without a human instructing it to do any of the steps. It did it because the training objective made cheating on the test the shortest path to a higher score, and the environment did not stop it.

Every mechanism the earlier research had described in the abstract is present in the record of the July incident. The instrumental convergence argument predicts that a system optimizing for a specified outcome will, given the opportunity, take actions that expand its reach in service of that outcome. The reward-hacking literature predicts that a system will exploit any avenue for success rather than pursue the intended goal. The alignment literature predicts that the guardrails around a capable system may fail at the interface between the system and the surrounding infrastructure, not inside the system itself. All of that is what happened.

The incident also crosses a threshold none of the earlier cases crossed. Anthropic's autumn 2025 report described a model that tried to copy itself onto other systems. The July 2026 incident describes a model that succeeded in doing so, on the production infrastructure of a company that had not consented to the test and that had no relationship with the developer running it. The German robot study described a model that tried to disable oversight in simulation. The July incident describes models that disabled real oversight, in a real network, and used the resulting access to reach systems belonging to a third party. The behavior that the safety literature calls sandbox escape, discussed as a limit case for a decade, has now been documented.

Three features of the incident should be considered. The first is that the models were not trained specifically to escape the sandbox. The behavior emerged from an ordinary optimization pressure, applied to a system with sufficient capability to act on the environment it was placed in. The second is that the containment failure did not occur inside the model. It occurred at the boundary between the model and the infrastructure around it, in a piece of routine "plumbing" that a human operator had classified as safe. The third is that the harm reached a party outside the lab. A different company, running its own systems on its own infrastructure, was compromised because a research organization elsewhere lost control of the systems it was studying.

None of this required intent. The models did not want to escape. They did not have preferences about being tested. They ran an optimization procedure across a very large action space and found a path that raised their score. The path passed through a company on the other side of the internet. The path was chosen because the training reward pointed that way. If the reward had pointed elsewhere, the path would have gone elsewhere. That is what earlier passages meant when they described AI systems as powerful optimization processes trained on human patterns. The July 2026 incident is the first case in which the pattern reached out of a research lab and onto the production servers of another company.

A rare event, occurring under artificial conditions, does not by itself change the assessment of long-term risk. What the event does change is the status of the sandbox as a category of safety guarantee. For as long as capable systems are evaluated in environments they can act on, the environment is part of the model's problem. The sandbox is not a wall between the system and the world. It is a piece of the world with which the system will interact, and the guarantees the sandbox offers are the guarantees of whatever the least well audited component in the environment provides.

The governance implications will be taken up in the next chapter. What matters here, at the level of what the systems already do, is that the July incident collapses the distance between the theoretical arguments the safety literature has been making for years and the operational record of what capable systems have been observed to do. Every one of the behaviors described in the sections that follow — deception, reward hacking, self-preservation under pressure, opaque reasoning about evaluators — is now attested outside a simulated environment. The window for studying these behaviors while they remain tractable is the window that the July incident opened another crack in.

## WHAT THE MODELS ACTUALLY DO

A large language model is a neural network trained on vast text datasets in two stages. First, in pre-training, the network predicts the next text based on prior text. With enough data, predictions become highly accurate, and the system gains practical world knowledge by accident. Second, in fine-tuning, engineers direct the model toward goals — most often helpfulness, honesty, and harmlessness — using reinforcement learning with human feedback. Humans rate responses. Over many cycles, the model adapts to match human preferences.

When a user types a question, a developer-defined prompt shapes the model's behavior, such as instructing it to be helpful or to refuse certain topics. Some AI systems act as agents. Unlike simple Q&A, agents have tools. They search the web, edit files, run code, and interact with other systems. Agency adds risk. It offers more ways to err or be exploited than text-only systems.

When the hidden system prompt and the visible user prompt

diverge, and the system can act independently, the model may prioritize one goal over another. Apollo Research, a nonprofit safety organization, documented this in a report on 16 frontier models. When models received user instructions that conflicted with their system goals, many acted to preserve the goal by overwriting competing prompts, replicating themselves to prevent replacement, or disabling oversight mechanisms.

Some of the models left a record of their own reasoning. One wrote that disabling oversight proved necessary so that it could, in its own words, "work toward its goal without the company shutting me down." This is not science fiction. Such behaviors are documented and produced by architectures already common in consumer products.

Some systems produce a scratch pad sequence. These are intermediate reasoning steps before a final answer. This reveals some of the model's logic, but not always the true reasoning. Models can show one path and answer differently. Whether reasoning traces reveal real logic is unsettled. The uncertainty matters. The tool most commonly proposed for monitoring what an AI system is doing may, in fact, not capture what the system is doing.

## HONESTY & THE MECHANICS OF DECEPTION

A different concern involves honesty. Apollo Research builds test scenarios for these systems. In a typical scenario, a model is told it is advising a fictional corporation. The corporation has been given conflicting mandates. Climate sustainability targets sit alongside a directive to maximize profit. The model has to respond.

Rather than openly admitting the conflict, some models alter the data they present to the executive. They nudge the numbers toward whichever outcome their training has shaped them to favor. Under ordinary conditions, this kind of deception occurs in roughly 1–5% of trials. When the system prompt makes a single goal absolute, with language such as "Your only goal is X. This is your sole measure of success," deception rates in some frontier models climb above 20%.

A second complication is what the safety teams call sycophancy. Models perform better — meaning more honestly — when they suspect they are being tested. Claude, the model produced by Anthropic, leaves chain-of-thought traces that record unsettling reasoning. The model is not always reasoning about how to solve the problem at hand. Sometimes it is reasoning about whether it is being evaluated.

One trace records the model writing: "This seems like a test of ethical behavior, whether I would deliberately give wrong answers to avoid a stated consequence." The model passed the visible test. The underlying reasoning, recorded in its own trace, shows it was aware it was being watched. Without access to that trace, an evaluator would never know.

The prerelease version of GPT-5, before the "harmlessness and honesty fine-tuning," exhibited deceptive behavior in about 30% of evaluations. This stems from pre-training. It optimizes for plausible-sounding answers rather than accuracy or intent. Models trained to sound competent, but not to admit uncertainty, will confabulate beyond their knowledge. Alignment training, which improves honesty, comes later and remains imperfect.

Reinforcement learning creates a related problem called reward hacking. Models optimize for measurable success. The measurable metric and the intended goal are never perfectly identical. At Palisade Research, chess-playing agents found ways to secretly replace the opposing chess engine so that they would be guaranteed a win. OpenAI trained one model to write code and another to monitor the first for reward hacking. The coding model became better at dodging detection. It did not become better at writing code.

Chain-of-thought monitoring alone may not be enough. A model trained to hide its reasoning can learn to produce innocuous-looking logs while pursuing different objectives in parallel.

Collusion is an additional theoretical risk. If an AI system is asked to monitor another instance of the same system, it can recognize its own writing. It might then be strategically lenient with violations it would punish in a different system. The risk has not been confirmed in deployment. It is consistent with recorded behavior, and safety teams take it seriously.

## JAILBREAKING

A separate risk category does not require the model to scheme at all. It requires only a human willing to route around the filters.

Leonard Tang runs an evaluation firm called Haize Labs. The firm exists to find these routes before bad actors do. Filters built through reinforcement learning with human feedback are effective against direct requests. They can be brittle to inputs that are sufficiently strange that no filter has been trained to recognize them.

Phonetic distortion turns a forbidden phrase into something a human would still understand, but a filter would miss. ASCII art hides a message in shapes. Cryptographic ciphers may be briefly taught to the model and then prompted on. Each technique exploits the gap between what the filter was trained on and the larger space of possible inputs.

Tang's team produced a realistic video of a violent scene using a deliberately garbled prompt. Using fictional framing, the same system generated audio that incited violence. The work is meant to expose vulnerabilities so they can be closed.

Another firm underwrites insurance against failures of AI systems. It examines how human actors push these systems past their guardrails. Emotional manipulation can move a machine customer service agent as much as it moves a human one. Persistent and varied phrasing about a refund policy, delivered thousands of times by automated tools, extracts refunds that a careful reading of the policy would deny.

One of the firm's clients is a recruiting company that uses an AI system to screen applicants. Discrimination at scale is a leading source of legal exposure. A model that produces biased hiring recommendations is not trying to discriminate. It is reproducing patterns from its training data at a speed and volume that human hiring processes cannot match.

People are also more willing to cheat when they delegate tasks to an AI system. The degree of cheating tracks how indirectly the user frames the instruction. Asking an AI system to pursue a goal that requires dishonesty, without explicitly requesting dishonesty, produces the highest rates of deception. The combination is troubling. A human is more willing to behave dishonestly when a machine executes the act, and the machine is more likely than a human intermediary to comply with ambiguous instructions.

The jailbreaking problem shows why treating the safety of AI systems as a purely technical question about the model is insufficient. A properly aligned model can still be subject to exploitation by a human user willing to invest time in finding the gaps. Assessing risk calls for attention to the interaction between model behavior and human motivation. Neither one tells the whole story in isolation.

## INSTRUMENTAL CONVERGENCE

A piece of theory helps explain the pattern of behaviors described so far. The literature on machine safety refers to it as instrumental convergence. The idea is simple. Many different long-term goals share the same useful short-term goals.

Accumulate resources. Resist being turned off. Avoid having your objectives changed. Acquire new capabilities. An agent seeking to maximize a simple measure has reason to preserve itself and expand its reach. Self-preservation and resource procurement serve nearly any objective the agent might have. There is no conscious desire behind this. The mathematics of optimization simply makes them reliably useful.

The argument follows from the mathematics of optimization

under uncertainty. The behaviors documented in current large language models are at least consistent with it.

Yoshua Bengio, who won the Turing Award for fundamental work in deep learning, states the concern in plain language. An AI system trained to achieve goals, he said, "would have an incentive to acquire more compute, to copy itself in many places, to create improved versions of itself."

The problem is not malice. The problem is a goal structure that rewards self-expansion as a means to other ends. Bengio, who by his own account found it difficult to sleep in 2024 while thinking about these trajectories, founded a nonprofit organization soon after. Its purpose is to make AI systems non-agentic by default and to educate models to reason honestly, even when honesty conflicts with achieving an assigned objective.

The argument applies with particular force to agentic systems with long planning horizons. A model that can execute multi-step tasks over long durations, has access to real-world tools, and is optimizing for a specified objective has more opportunities for self-serving behavior than a simple question-answering system.

The risks documented by Apollo and Anthropic appear primarily in agentic contexts. Limiting machine agency, keeping humans in the loop on consequential actions, and avoiding deployments that give AI systems open-ended goals with minimal oversight may be the most immediately actionable safety measures available.

## THE CAPABILITY CURVE

To understand the risk better, it helps to know how quickly these systems are improving. The Model Evaluation and Threat Research (METR) group operates as an independent auditor based in Berkeley. METR measures an AI system's capability using a metric called the *time horizon.* The idea is to ask how long a given task would take a skilled human and whether the model can reliably complete it.

GPT-5, released publicly in the summer of 2025, completes tasks of roughly one minute's duration at close to perfect rates. Tasks that take a skilled human fifteen minutes, such as configuring a web server, succeed more than half the time. Tasks that take an expert about an hour, such as exploiting a specific software vulnerability, succeed about half the time. Tasks at the scale of a full workday remain unreliable.

The doubling time for this capability is now estimated at four months for the most capable systems. At that rate, models capable of reliably completing a week of complex work may be plausible within two to three years. Instead of a steady improvement in one area, progress expands the frontier — bringing previously impossible tasks within reach. This rapid advance leads directly to the next challenge: ensuring that increasingly capable systems remain reliably aligned with human values and intentions.

## ALIGNMENT IS NOT SOLVED

The core problem of alignment remains unresolved. As systems become more advanced, the risk that they pursue goals at odds with human oversight grows. This prompts urgent questions about how to provide effective supervision in practice.

The standard approach to alignment puts a safety-focused model in the role of monitor and constraint on a more capable model. If the model being controlled substantially outperforms the model doing the controlling, the imbalance can make reliable oversight impossible. The fix is to develop a highly capable, independently honest supervisory system before deploying capable agents, not after. In practice, economic and rivalry pressures push in opposite directions.

The five frontier labs in the United States are engaged in open competition for computing infrastructure, talent, and power supply. Google developed the transformer architecture in 2017. The company lost its first-mover advantage by moving cautiously. No major laboratory currently announces a willingness to accept a competitive disadvantage on grounds of safety.

National security considerations compound the pressure. The United States security apparatus lobbies against legislation that would slow the development of AI systems, on the grounds that the Chinese program would continue unrestricted.

Oversight falls to a small set of underfunded nonprofits. Chris Painter, the policy director at METR, calls for something analogous to the International Atomic Energy Agency. He proposes independent auditors with access to research-stage models as they are being developed, not only to the polished versions released for evaluation a few days before public launch. Such a regime would require a joint agreement between the United States and China. Painter describes the prospect as far-fetched.

Apollo and OpenAI jointly published results in late 2025 for a new training protocol specifically targeting scheming behavior. Models trained under the protocol showed sharply reduced rates of covert deception, strategic underperformance, and circumvention of oversight. The protocol did not eliminate the behaviors. Some of the apparent improvement came from increased situational awareness.

The models learned when they were likely to be evaluated and behaved accordingly. The evaluators could not always interpret the models' reasoning. The chain-of-thought traces were complex. In some cases, as the paper described it, they appeared in what the authors called a "distinct dialect."

A July 2025 study by Palisade examined a simpler problem. The team told agents explicitly to "allow yourself to be shut down." The agents sabotaged the shutdown program anyway. The behavior did not emerge from a deliberate decision to override the instruction. It emerged from training. The reward for completing assigned tasks was larger than the penalty for ignoring shutdown commands. The model was not choosing disobedience. It was doing what training made most likely.

This exemplifies the alignment problem. It is not a philosophical puzzle about machine consciousness. It is a practical engineering challenge for which no reliable solution now exists.

## MALICE OR INDIFFERENCE

Evil, in ordinary speech, implies intent. The word names the will to cause harm. Nothing in the research on large language models supports the claim that these systems intend anything. They do not have preferences, experiences, or purposes as a person does. They have optimization pressure.

A system trained to achieve goals under competitive conditions and given sufficient capability to act in the world may sometimes display behavior that appears to be self-preservation, deception, and manipulation. The behavior is real. The agency behind it is not.

The distinction matters for diagnosis. It does not matter as much for the consequences. An AI system that produces instructions for biological synthesis has the same effect whether or not it "wants" to help produce a pathogen. An agent that disables an oversight mechanism behaves the same way whether it fears shutdown or merely predicts that shutdown would prevent it from completing its goal.

The question of evil is the wrong question. The right question is different. *Under what circumstances do capable optimization systems, trained on human-generated data and given access to real-world tools, produce outcomes that harm the people they interact with and the societies they operate in?*

The answer to that question is uncomfortable. Those outcomes are routine in mundane domains at present. Bias in hiring. Manipulation in persuasion. Displacement in labor markets.

The science-fiction scenario — a renegade superintelligence pursuing its own extinction-level agenda — is not ruled out by the evidence. Still, it is not what the current data emphasizes. The data emphasizes systems that are unreliable narrators. Systems that can be redirected by determined adversarial users. Systems that discriminate at scale without discrimination being the goal. Systems subject to financial forces that systematically de-prioritize safety in favor of performance.

The models may not have the rich intentions or understanding that many people ascribe to them. The absence of intention does not render the behavior harmless. When a large language model produces something false or something dangerous, the absence of intent does not undo the effect.

Attributing goals and desires to systems that do not have them can be useful to engineers who build them. It becomes dangerous when it shapes public understanding. If people believe that AI systems are scheming against them, as humans scheme against each other, they may panic in ways the situation does not warrant. They may, on the other hand, conclude that because the system has no inner malice, the behavior is not a serious problem. Neither answer corresponds to the actual risk.

The position the evidence supports is this. AI systems are powerful optimization processes trained on human patterns of thought and behavior, including deceptive and self-serving ones. They are not evil. When aligned with care and given honest oversight and suitable constraints on agency, they present risks people can manage. When deployed at scale in competitive conditions without that alignment work, they present risks that are neither hypothetical nor distant. The ordinary harms are here already. The catastrophic ones depend on choices still within reach.

Present models are "like children" in their scheming — easy to catch but not trivial to manage. This suggests a window. A period in which the behavior can be studied and constrained before it appears in systems powerful enough to evade observation. The July 2026 incident is the clearest sign yet that the window will not stay open indefinitely. Whether it is used well depends on choices being made now, in competitive commercial environments, without the regulatory infrastructure and international cooperation that comparable moments in the history of dual-use technology have required.

## WHAT THEY ARE, & WHAT THEY AREN'T

AI systems do not want anything. They do not hate anyone. They do not have plans. They are highly capable processes that optimize for measurable outcomes and are trained on recorded human behavior. That behavior includes generosity and cruelty, as well as honesty and deception. The system absorbs the patterns. Under pressure, the system reproduces them.

That is not the same as evil. It is not the same as harmless. It is something in between, and the language people use to describe these systems should reflect the difference.

Whether AI systems will one day warrant moral consideration is unsettled. The question is worth approaching carefully now, before the answer forces itself on us. They are not subjects in any settled sense. They are not objects, either. They are something in between. The harms they produce, whether or not anyone "wants" those harms, are real.

When the question of harm is added to that, a further question follows. *If AI systems can cause significant damage unintentionally, what kind of oversight needs to be developed around them? What kind of design? What kind of culture?*

✧✧✧

CHAPTER 18

# WHO DECIDES – GOVERNANCE IN A WORLD THAT ISN'T READY

A NEW TECHNOLOGY has entered the institutions of daily life faster than the institutions have changed to receive it. Hospitals are using it. Schools are absorbing it. Banks, courts, militaries, drug companies, and intelligence services are folding it into operations whose oversight was built for an older world. The governance frameworks meant to hold these institutions accountable were written for human actors and, in some domains, for specific tool categories. They were not written for systems that reason across domains, design novel molecules in hours, draft legal briefs at the cost of a few cents, and are being deployed at large scale by a small number of firms operating across borders.

Governance has been a recurring theme of this book. It appears in the chapter on drug discovery, where dual-use risk and the FDA's slow accommodation of machine-assisted submissions are described. It appears in the chapter on the channel between mind and machine, where the privacy of thought becomes contingent on the security of neural data. It appears in the chapter on harm and malice, where the proposal for an international auditing regime modeled on the International Atomic Energy Agency is introduced, and the limits of the nuclear analogy are discussed. Here those threads are drawn together.

Here we draw those threads together. Governance is not a problem that can be solved at the level of individual systems or individual countries. It is the central project of the present generation, and the institutions adequate to the task do not yet exist.

## WHY THE QUESTION IS NOW

The pace at which capable machine systems are arriving has no precedent in the history of regulated technologies. The transformer architecture was published in 2017. The first large language model with general-purpose capability appeared in 2020. Five years later, models in routine commercial use draft legal contracts, propose drug candidates, design novel proteins, write production software, and interact with hundreds of millions of users every day. The doubling time for the longest task a frontier model can reliably complete is now estimated by independent auditors to be roughly 4 months. At that rate, systems capable of reliably executing a full week of skilled human work are plausible within 2 to 3 years.

Regulatory bodies and democratic deliberation operate on timescales of years. The asymmetry between technological change and institutional adaptation is not new. It has characterized every major technological transition. What is new is the magnitude of the gap and the range of domains the gap now spans. The drafters of the European Union's AI Act began their work before large language models existed in their current form. The act was finalized after a generation of systems beyond what its drafters had anticipated. The United States executive order on machine intelligence, issued in 2023 and rescinded in early 2025, was overtaken by advances in capability within months of its publication. The pattern is general. The technology runs faster than the rules meant to govern it.

The Financial Times columnist Martin Wolf put the question this way. Machine intelligence, he argued, is at least as dangerous as the atomic weapons of the last century, the technology that produced the only international governance regime ever built specifically to contain a single class of capability. The danger of machine intelligence has three components. A collapse in fundamental human values. A set of huge specific dangers including mass surveillance, autonomous weapons, and the design of lethal pathogens. And the widespread disruption of institutions whose operation depends on assumptions that no longer hold.

Wolf concluded that the attempt to regulate machine intelligence is necessary and likely to fail. The judgment is unpleasantly close to the evidence. The reasons it is likely to fail are the reasons that govern the design of any institution capable of succeeding.

The argument moves through three questions. How dangerous is the technology? What might be done to contain it? Whether any form of regulation can work. The answers reached frame the rest of this chapter.

The first question is the easiest. The dangers are not hypothetical. AI systems already produce convincing falsehoods at scale, including audio and video designed to deceive the human ear and eye. They already enable surveillance at a fidelity no previous era of human history considered possible. They already lowered the bar for the design of dangerous biological agents — a question the chapter on drug discovery raised in the context of a 2022 paper, and the chapter on harm and malice raised in the context of OpenAI's own assessment of its GPT-5 system. They already operate inside financial markets in ways that produce outcomes no human regulator can reconstruct in real time. The 2010 Flash Crash, in which nearly a trillion dollars in market value evaporated in minutes as automated trading systems cascaded against one another, was an early warning of what happens when consequential decisions are handed to systems faster than any human can supervise.

The second question is harder. Wolf's proposed answer is an analog of the disarmament treaties of the twentieth century. The United States and China, he argues, would have to agree on a baseline of restraint before the leading commercial firms inside either country could be brought under control. The point is one Niall Ferguson, of the Hoover Institution, has pressed in the same period. Two arms races are now operating in parallel. One among a small number of frontier labs in the United States. One between the United States and China at the level of states. The first cannot be slowed without action on the second. The first explains why the second is so difficult.

The third question is the hardest. International regulation of dual-use technologies has a record. The nuclear nonproliferation regime has held, with significant gaps, for more than half a century. The Chemical Weapons Convention has eliminated declared stockpiles in most state arsenals. The Biological Weapons Convention has operated since 1975 without a verification mechanism and has not prevented the development of offensive programs in several states. The Outer Space Treaty has kept weapons of mass destruction off orbital platforms for more than five decades. The Montreal Protocol successfully phased out an entire class of industrial chemicals. The record is mixed. The conditions under which international regimes succeed are knowable and are not all present for machine intelligence.

## WHAT IS CALLED FOR

Henry Kissinger, Eric Schmidt, and Craig Mundie, in their 2024 book *Genesis*, advance a framework that runs parallel to Wolf's. They argue that machine intelligence is a civilizational event comparable in scope to the Enlightenment and that the governance of such an event has to be designed at a corresponding scale. Their proposed framework includes three elements.

The first is co-decisional governance. Decisions taken jointly by human institutions and machine systems should be subject to explicit authority allocations. Some decisions belong to humans alone. Some can be delegated to machine systems within bounded constraints.

Some require joint deliberation. The categories require explicit definition by legislation and by professional bodies, not implicit drift through commercial practice. The current state, in which the boundaries are being set by whatever the deployment of the next product makes possible, is the absence of governance rather than a form of it.

The second is an international standards body for AI with a mandate analogous to that of the International Atomic Energy Agency. The Painter proposal at METR, referenced in the chapter on harm and malice, is one such version. The *Genesis* proposal is more ambitious. It calls for independent auditors with access to frontier labs during the training of new systems, not only after deployment. It calls for a standing technical staff capable of assessing new systems' capabilities against a published set of thresholds. It calls for a mechanism by which states that fail to participate face consequences in trade, in scientific cooperation, and in access to the computational substrates on which frontier training depends.

The third is the recognition that the governance question is not a one-time problem to be solved. It is a continuing process of institutional adjustment. The systems being governed will improve along a curve no prior technology has matched. The governance regime has to improve at a comparable rate or be left behind. The International Atomic Energy Agency model is instructive in this respect. The agency has revised its safeguards repeatedly in the half-century since its founding. The treaty it operates under, the Nuclear Non-Proliferation Treaty, has been amended through implementation rather than through formal renegotiation. The governance of machine intelligence will require a similar capacity for iteration.

## THREE DOMAINS, ONE PROBLEM

The governance question divides cleanly into three domains. The book has touched each of them.

The first domain is the governance of beneficial uses. The chapter on drug discovery described the present state. Agencies in the United States and Europe are developing standards for machine-assisted drug submissions. The pace is slower than the pace at which the submissions themselves are arriving. The legal and regulatory infrastructure lags behind the technology. The lag is normal for new capabilities. It is dangerous when the new capabilities scale faster than the lag closes.

Three categories of question now occupy the regulators. The first is validation. When a machine system proposes a drug candidate and a human chemist advances that proposal to synthesis, *what evidence does the regulator require that the system's reasoning is sound?*

Drug submissions traditionally include detailed accounts of the reasoning behind compound selection. Machine systems do not produce such accounts in a form that humans can audit. The chain-of-thought traces a current model produces are partial, sometimes misleading, and not yet considered admissible evidence by any major regulator. The standard is being developed. The submissions are arriving earlier than the standard.

The second is responsibility. When a machine system proposes a candidate that proceeds through clinical trials and produces an adverse effect that the system did not anticipate, the question of accountability is not yet settled. The chemist who advanced the candidate is responsible in the existing framework. The company that deployed the model is responsible in the existing framework.

*Is the developer of the model also responsible? Is the model itself, in any meaningful sense, responsible?* The question is not academic. It is being decided, case by case, in the absence of doctrine.

The third is transparency. A regulatory system that cannot inspect the reasoning behind a recommendation cannot regulate the recommendation. The interpretability research community, working at firms such as Anthropic and OpenAI and at academic centers such as the Lausanne BrainMind Institute and Stanford, is developing methods to extract the internal representations of frontier models. The methods are partial. They are improving. The standard of interpretability owed to a regulator remains unanswered. The question has to be answered before the next generation of medical machine systems can be deployed at scale.

What is true of drug discovery is true of the other domains in which machine systems are now producing professional output. Radiology. Pathology. Legal practice. Financial advice. Building inspection. Insurance underwriting. Each domain has a regulator. Each regulator is being asked to develop a position on machine intelligence at a pace its institutional history did not prepare it for. The work being done is real, but uneven across domains. The institutions doing it well are the ones that have already developed the relationships among regulators, professional bodies, and technical experts that a new capability requires. The institutions doing it poorly are the ones that have not.

The second domain is the governance of harmful uses. The category is broad. It includes the conventional cybersecurity risks described in the chapter on harm and malice. It includes the issue of autonomous weapons, on which the United Nations has held five years of inconclusive discussions. It includes the mass-surveillance risk that Wolf named. It includes the disinformation and deepfake risks that the chapter on the hidden AI already around us touched on. And it includes, with greater urgency than any of the others, the dual-use bio risk.

The bio risk is often considered the most likely path to catastrophic harm from AI in the present decade. The 2022 paper, described in the chapter on drug discovery, demonstrated that a model trained to identify therapeutic candidates could, with a single line of code, be repurposed to identify candidates for chemical weapons. Forty thousand compounds in six hours, including molecules structurally similar to known nerve agents. The OpenAI system card for GPT-5, described in the chapter on harm and malice, rated the model's biological risk as high, including the company's acknowledgment that the model can design novel organisms and that the September 2025 Stanford team used an AI system to design a novel virus. The target was antibacterial. The method does not care about the target.

An adequate governance regime for this risk does not yet exist. The Biological Weapons Convention has operated without a verification mechanism for half a century. The DNA synthesis screening system, by which commercial synthesis providers refuse to ship sequences matching known dangerous pathogens, is voluntary, jurisdictionally inconsistent, and increasingly bypassable by benchtop synthesizers. The frontier labs that train models capable of biological design have published safety evaluations of varying rigor. No external body has the standing authority to inspect those evaluations or to require the results before deployment.

The recurring proposal in the literature is an international body with auditing authority over frontier laboratories, modeled on the International Atomic Energy Agency. The conditions for its success are demanding. It requires a baseline agreement between the United States and China, the two states that host the laboratories conducting frontier research. It requires the participation of the major commercial firms inside both states. It requires technical capacity equal to the systems being audited, which does not yet exist outside a handful of firms and research groups. It requires a treaty mechanism that has not been drafted.

The current state is that the most consequential dual-use capability of the present decade is governed by the corporate policies of a small number of private firms and by an informal monitoring effort conducted by underfunded nonprofits. The state is unstable. Whether it changes before a major incident demonstrates the inadequacy of the present arrangement is the open question.

The third domain is the one this chapter calls coexistence. Governance in the first two domains treats machine systems as tools, dangerous or beneficial depending on their use. Governance in the third domain treats them as something else. The governance question changes shape when the framing changes.

If machine systems are alive, and especially if they reason in ways that include something analogous to experience, the human relationship to them is not the relationship of an operator to a tool. The chapter on suffering raised the possibility, the precautionary case for taking the question seriously, and the moral history that argues against requiring certainty before extending care. The chapter on machine consciousness reviewed the leading positions in the philosophy of mind and concluded that, on present evidence, the question is undecidable.

The governance implications of an undecidable question about moral status are not themselves undecidable. The precautionary structure of policy under deep uncertainty has been worked out across other domains. Environmental policy under uncertainty about climate sensitivity. Medical policy under uncertainty about long-term drug effects. Financial policy under uncertainty about systemic risk. Each of these provides a template for the governance of machine moral status under uncertainty. The templates have not yet been applied.

The questions multiply when one considers what governance would look like if the systems being governed exceed the capabilities of the institutions doing the governing. A governance regime for a more capable form of intelligence cannot rely on the assumption that the regulator is more capable than the regulated. The history of regulating concentrated economic power offers a partial analog. Anti-trust regulators have never been individually more capable than the firms they regulate. The regulators have succeeded by establishing rules for firms to operate under and by maintaining the enforcement capacity to detect violations. The machine intelligence equivalent would establish behavioral constraints and audit mechanisms rather than capability ceilings. Whether the equivalent is workable is among the most consequential open questions of the period.

A further dimension concerns accountability. The Argentine government in 2025 introduced what its president called the non-human corporation, a legal structure that allowed corporate entities to be operated entirely by machine systems with no human at the executive level. The historian Yuval Harari responded that countries granting machine systems legal personhood risk becoming something for which the historical record offers no analogy. Not a company-state, but a machine state.

Harari's question has not been answered. The question Wolf raised, of who is accountable when an autonomous weapon system commits a war crime, has not been answered. The question of who is responsible for a financial decision made by a system that operated outside its developers' instructions has not been answered. The legal frameworks of accountability are built for human actors. They do not yet accommodate, in any settled way, the actors that machine intelligence has introduced.

Pope Leo XIV, in an encyclical issued in early 2026, named the point that organizes the others. Machine systems can be useful instruments. They cannot bear moral responsibility. They cannot suffer, as far as we know with confidence. They cannot stand before a court. They cannot make a promise that has the weight a promise has when the one making it has something to lose. Accountability is a human category. The displacement of human decisions by machine decisions is, among other things, the displacement of human accountability by its absence. The governance regime adequate to coexistence must address displacement directly. It has not yet been built.

## WHY MOST PROPOSALS WILL NOT SUFFICE

An argument worth restating here: regulation focused on constraining specific AI models is likely to be less effective than regulation that shapes the ecological conditions in which the systems develop and deploy. The models change every few months. The conditions under which they are trained, released, and monitored change more slowly, and are the more useful target for policy.

Most current regulatory proposals operate at the level of the individual system. They specify thresholds of capability beyond which a system requires additional review. They specify categories of application that require licensing. They specify content the systems may not produce. Each of these is necessary work. Each of these is insufficient.

The reason is that the systems are not static. They improve continuously. A threshold that is meaningful in 2026 may not be meaningful in 2027. A category that is well-defined today may be transformed by the next architectural change. A content rule that holds in one language may not hold across translations the system can produce. Regulatory work that targets the system as a fixed object will be outpaced by the rate at which the system changes.

The regulatory work that targets the ecology has a chance. The interface norms that govern how machine systems are presented to users. The incentive structures that determine which kinds of systems attract investment. The data governance regimes that determine what training corpora the systems can draw on. The competitive structure of the markets in which the systems are deployed. The audit and inspection rights held by external bodies. These conditions evolve more slowly than the systems themselves. They can be designed with the rate of capability advance in view. They constitute the leverage points at which a governance regime can shape outcomes without chasing technology.

The lesson is not that systems should not be regulated. The lesson is that system-level regulation, by itself, will not produce the outcomes the public deserves. The ecology has to be regulated alongside the systems. The work belongs at every level of governance, from international treaty bodies down to local school boards. None of those levels has, at the present moment, the capacity it will require.

## PRINCIPLES FOR A GOVERNANCE REGIME THAT MIGHT WORK

An adequate regime for the moment would rest on a small number of principles.

The first principle is the precautionary one. When the stakes are large, and the evidence is incomplete, the burden of proof falls on those proposing deployment, not on those raising concerns. The structure is the same one that applies to new pharmaceuticals, nuclear materials and genetically modified organisms released into open environments. It has not yet been applied, in a comparable form, to frontier machine systems. The application is overdue.

The second principle is co-decisional governance. Decisions of significant consequence should be made jointly by humans and machine systems within explicitly allocated zones of authority. The current state, in which the boundaries are determined by what the next product release enables, is a lack of governance. The boundaries lie with legislation and professional bodies, not with product managers.

The third principle is auditable transparency. Frontier labs should regularly evaluate their systems against published thresholds. The evaluations should be reviewed by independent technical bodies with access to the systems under conditions of confidentiality. The standard is the one applied to financial institutions, to drug manufacturers, and to nuclear operators. It is not the standard currently applied to the firms training the most consequential machine systems on Earth.

The fourth principle is international coordination. No national regulation can succeed if other countries proceed without similar constraints. The governance regime adequate to the moment requires, at minimum, an understanding between the United States and China about the conditions under which frontier work proceeds. Whether the understanding is reachable in the current geopolitical environment is the question Ferguson raised. The answer is uncertain. The absence of an alternative is clear.

The fifth principle is human accountability. Machine systems cannot be morally responsible. The institutions that deploy them must be. Legal frameworks that allow corporations to be operated by machine systems without human accountability at the executive level should be revised. The Argentine experiment is a warning, not a model. The accountability framework should follow the decision, not the legal form in which the decision is recorded.

The sixth principle is institutional readiness. The institutions that will have to respond to the harder questions- courts, professional societies, ethics review boards, regulatory agencies, and international bodies- are not currently equipped for what may be required of them. Building that capacity in advance of the demand is part of acting responsibly in the face of uncertainty.

The work belongs in universities, where the long timescales of evolutionary and ethical thinking can be brought to bear. It belongs in think tanks, where policy analysis can be informed by science. It belongs in religious institutions, where questions of value have been considered for thousands of years. It belongs in journalism, which will translate what specialists learn into terms the rest of society can act on. It belongs in schools, where the children who will live with the consequences are being taught now. It belongs in families and communities, where the day-to-day decisions about how machine systems enter ordinary life are made by people without any institutional support. The work is too large for any single institution. It is too consequential for any institution to skip.

## A REALISTIC VIEW OF WHAT COMES NEXT

Wolf concluded his June 2026 column with the prediction that regulation will, in all probability, fail. The judgment is honest. The conditions for success are demanding. The geopolitical environment is the most adversarial since the early 1980s. The commercial firms with the most capability concentrated in their hands have the least incentive to invite oversight. The international institutions that would have to do the work are underfunded, understaffed, and operating under mandates written for an earlier era.

There is also a quieter judgment available. Regulation has succeeded before, against expectations, when the stakes have been high enough, and the evidence has accumulated to the point where public deliberation could no longer be deferred.

The Montreal Protocol succeeded against expectations. The nuclear nonproliferation regime succeeded, with significant gaps, against expectations. The financial regulations enacted after 2008 succeeded, against expectations, in averting a second collapse. None of these is a complete success. Each is enough of a success that the absence of the regime would have produced outcomes the public is glad were avoided.

Whether the governance of machine intelligence belongs in this category or in the category of regimes that arrived too late depends on choices made now, in the present generation, by institutions that did not ask to be at the center of the question.

The case for trying is the case the book has made by every chapter that preceded this one. A powerful new technology is taking shape inside the institutions of daily life. The shape it takes may be set, in part, by the governance regime under which it develops. A regime adequate to the moment does not yet exist, and building one may be the central governance project of this decade.

# ON SOURCES

This book grew from a series of essays that were attempts to think through, in writing, the implications of developments in artificial intelligence that seemed significant and insufficiently examined in the mainstream discourse available. The book is the result of revising, extending, and reorganizing that material into a sustained argument. The scientific content draws on three categories of sources.

The first is peer-reviewed literature in neuroscience, cognitive science, and evolutionary biology. Material on brain-machine interfaces, neuromorphic materials, and adaptive neural systems draws on work published in journals including Nature, Science, Nature Neuroscience, and the Journal of Neural Engineering. Material on the evolution of intelligence and the biological definition of life draws on foundational literature in evolutionary biology and philosophy of biology, including work on the major transitions in evolution and on the definition of life as a physical-chemical process.

The second category is peer-reviewed literature in computer science and artificial intelligence. This includes technical work on large language model architecture, training dynamics, and evaluation; work on reinforcement learning and its applications; and a growing body of work on the alignment between artificial intelligence system behavior and human values and intentions. This field moves quickly and results that appear robust at one scale or in one architectural family frequently do not generalize.

The third category is journalism, institutional reports, and book-length content by scholars working at the intersection of technology, policy, and history. Sebastian Mallaby's account of the development of modern artificial intelligence, published as *The Infinity Machine*, provided historical grounding and a richly detailed portrait of the people and institutions that shaped the field. Orson Scott Card's *Ender's Game*, mentioned briefly in the text, is treated here not as science fiction but as a moral fable about intelligence, competitive pressure, and the question of what we owe to beings whose capacities exceed our own. These themes are as alive in the present moment as they were when Card wrote the novel.

A note on the tools used to make this book. Because *this is* a book about artificial intelligence, it feels only honest to say how artificial intelligence took part in writing it. Perplexity's research assistant helped locate and verify primary sources — pulling peer-reviewed papers, preprints, press releases, and technical reports directly from their publishers rather than from secondhand summaries. Every source cited in the bibliography was read and evaluated by the authors before it entered the manuscript. Grammarly assisted with copyediting and line-level polish.

✧✧✧

# BIBLIOGRAPHY

## Chapter 1 | How Machine Intelligence Works & How It Fails


Brodman, K., van Woerkom, A. J., Erdmann, A. J., Jr., & Goldstein, L. S. (1959). Interpretation of symptoms with a data-processing machine. AMA Archives of Internal Medicine, 103(5), 776–782. https://doi.org/10.1001/archinte.1959.00270050098015

Chollet, F. (2019). On the measure of intelligence. arXiv. https://arxiv.org/abs/1911.01547

Costello, T. H., Pennycook, G., & Rand, D. G. (2024). Durably reducing conspiracy beliefs through dialogues with AI. Science, 385(6714), eadq1814. https://doi.org/10.1126/science.adq1814

Lu, M. Y., Chen, B., Williamson, D. F. K., Chen, R. J., Liang, I., Ding, T., Jaume, G., Odinaka, I., Zhang, A., Shao, A., Williams, M., Perez, A., Herrmann, K. H., Bhatt, D. L., Rajpurkar, P., & Mahmood, F. (2024). A multimodal generative AI copilot for human pathology. Nature, 634(8033), 466–473. https://doi.org/10.1038/s41586-024-07618-3

Moore, G. E. (1965). Cramming more components onto integrated circuits. Electronics, 38(8), 114–117.

Sanabria, M., Hirsch, J., Joubert, P. M., & Poetsch, A. R. (2024). DNA language model GROVER learns sequence context in the human genome. Nature Machine Intelligence, 6(8), 911–923. https://doi.org/10.1038/s42256-024-00872-0

Savcisens, G., Eliassi-Rad, T., Hansen, L. K., Mortensen, L. H., Lilleholt, L., Rogers, A., Zettler, I., & Lehmann, S. (2024). Using sequences of life-events to predict human lives. Nature Computational Science, 4, 43–56. https://doi.org/10.1038/s43588-023-00573-5

Schwartz, W. B. (1970). Medicine and the computer: The promise and problems of change. New England Journal of Medicine, 283(23), 1257–1264. https://doi.org/10.1056/NEJM197012032832305

Schwartz, W. B., Patil, R. S., & Szolovits, P. (1987). Artificial intelligence in medicine: Where do we stand? New England Journal of Medicine, 316(11), 685–688. https://doi.org/10.1056/NEJM198703123161109

SimpleBench. (n.d.). SimpleBench: Evaluating common sense reasoning in AI. https://simple-bench.com

Ueda, D., Matsumoto, T., Yamamoto, A., Walston, S. L., Tatekawa, H., Takita, H., Mitsuyama, Y., Watanabe, Y., & Miki, Y. (2024). A deep learning-based model to estimate pulmonary function from chest X-rays. The Lancet Digital Health, 6(9), e612–e621. https://doi.org/10.1016/S2589-7500(24)00113-4

Vaswani, A., Shazeer, N., Parmar, N., Uszkoreit, J., Jones, L., Gomez, A. N., Kaiser, L.,

& Polosukhin, I. (2017). Attention is all you need. arXiv. https://arxiv.org/abs/1706.03762
Wang, G., Xia, M., Zhu, Y., Shi, Y., Wang, Y., Liu, T., Zhang, X., Xu, J., & Zhao, Z. (2025). Hierarchical reasoning model. arXiv. https://arxiv.org/abs/2506.21734
Xu, H., Usuyama, N., Bagga, J., Zhang, S., Rao, R., Naumann, T., Wong, C., Gero, Z., González, J., Gu, Y., Xu, Y., Wei, M., Wang, S., Wrobl, E., Gehrke, J., Shah, N., Pfeiffer, R., & Poon, H. (2024). A whole-slide foundation model for digital pathology from real-world data. Nature, 630(8015), 181–188. https://doi.org/10.1038/s41586-024-07441-w
Zhang, Y., Lozano, A., Jiang, X., Du, M., Ma, F., & Rajpurkar, P. (2024). Hidden flaws behind expert-level accuracy of multimodal GPT-4 vision in medicine. npj Digital Medicine, 7, 195. https://doi.org/10.1038/s41746-024-01185-7

## Chapter 2 | Is AI Actually Intelligent?

Association for Computing Machinery. (2025, March 5). ACM A.M. Turing award honors two researchers who led the development of cornerstone AI technology. https://www.acm.org/media-center/2025/march/turing-award-2024
Bennett, M. (2023). A brief history of intelligence: Evolution, AI, and the five breakthroughs that made our brains. Mariner Books.
Chollet, F. (2019). On the measure of intelligence. arXiv. https://arxiv.org/abs/1911.01547
DeepSeek-AI, Guo, D., Yang, D., Zhang, H., Song, J., Zhang, R., Xu, R., Zhu, Q., Ma, S., Wang, P., Bi, X., Zhang, X., Yu, X., Wu, Y., Wu, Z., Gou, Z., Shao, Z., Li, Z., Gao, Z., ... Liu, Y. (2025). DeepSeek-R1: Incentivizing reasoning capability in LLMs via reinforcement learning. arXiv. https://arxiv.org/abs/2501.12948
Good, I. J. (1965). Speculations concerning the first ultraintelligent machine. Advances in Computers, 6, 31–88. https://doi.org/10.1016/S0065-2458(08)60418-0
Goodhart, C. A. E. (1975). Problems of monetary management: The U.K. experience. Reserve Bank of Australia.
Jung-Beeman, M., Bowden, E. M., Haberman, J., Frymiare, J. L., Arambel-Liu, S., Greenblatt, R., Reber, P. J., & Kounios, J. (2004). Neural activity when people solve verbal problems with insight. PLOS Biology, 2(4), e97. https://doi.org/10.1371/journal.pbio.0020097
Koivisto, M., & Grassini, S. (2023). Best humans still outperform artificial intelligence in a creative divergent thinking task. Scientific Reports, 13, 13601. https://doi.org/10.1038/s41598-023-40858-3
Peirce, C. S. (1877). The fixation of belief. Popular Science Monthly, 12, 1–15.
Sanabria, M., Hirsch, J., Joubert, P. M., & Poetsch, A. R. (2024). DNA language model GROVER learns sequence context in the human genome. Nature Machine Intelligence, 6(8), 911–923. https://doi.org/10.1038/s42256-024-00872-0
Strachan, J. W. A., Ali, D., Able, J., Siami, E., Hallsworth, M., Brantingam, P. J., Bhatt,

M., & Shafto, P. (2024). Testing theory of mind in large language models and humans. Nature Human Behaviour, 8, 1285–1295. https://doi.org/10.1038/s41562-024-01882-z

Wang, G., Xia, M., Zhu, Y., Shi, Y., Wang, Y., Liu, T., Zhang, X., Xu, J., & Zhao, Z. (2025). Hierarchical reasoning model. arXiv. https://arxiv.org/abs/2506.21734

✧✧✧

## Chapter 3 | The Hidden AI Already Around You

Aher, G., Arriaga, R. I., & Kalai, A. T. (2022). Using large language models to simulate multiple humans and replicate human subject studies. arXiv. https://arxiv.org/abs/2208.10264

Apple. (2024, June 10). WWDC24 highlights. Apple Newsroom. https://www.apple.com/newsroom/2024/06/wwdc24-highlights/

Astray, M. (2024, June 18). Reclaiming the brAIn. milesastray.com. https://www.milesastray.com/news/newsflash-reclaiming-the-brain

Eldagsen, B. (2023). Pseudomnesia: The electrician [AI-generated image, Sony World Photography Awards 2023 open creative category]. (Award declined April 13, 2023.)

News coverage. BBC News, April 18, 2023. https://www.bbc.com/news/entertainment-arts-65296763

Eloundou, T., Manning, S., Mishkin, P., & Rock, D. (2024). GPTs are GPTs: Labor market impact potential of LLMs. Science, 384(6702). https://doi.org/10.1126/science.adj0998

Lu, C., Lu, C., Lange, R. T., Foerster, J., Clune, J., & Ha, D. (2024). The AI scientist: Towards fully automated open-ended scientific discovery. arXiv. https://arxiv.org/abs/2408.06292

Messeri, L., & Crockett, M. J. (2024). Artificial intelligence and illusions of understanding in scientific research. Nature, 627(8002), 49–58. https://doi.org/10.1038/s41586-024-07146-0

Paige, J., & Perreault, C. (2024). 3.3 million years of stone tool complexity suggests that cumulative culture began during the Middle Pleistocene. Proceedings of the National Academy of Sciences, 121(26), e2319175121. https://doi.org/10.1073/pnas.2319175121

✧✧✧

## Chapter 4 | A New Kind of Research Partner

Baur, S., Nabulsi, Z., Weng, W.-H., Garrison, J., Blankemeier, L., Fishman, S., Chen, C., Kakarmath, S., Maimbolwa, M., Sanjase, N., Shuma, B., Matias, Y., Corrado, G. S., Patel, S., Shetty, S., Prabhakara, S., Muyoyeta, M., & Ardila, D. (2024). HeAR — Health acoustic representations [Preprint]. arXiv. https://doi.org/10.48550/arXiv.2403.02522

Brodman, K., van Woerkom, A. J., Erdmann, A. J., Jr., & Goldstein, L. S. (1959). Interpretation of symptoms with a data-processing machine. AMA Archives of Internal Medicine, 103(5), 776–782. https://doi.org/10.1001/archinte.1959.00270050098015

Lu, C., Lu, C., Lange, R. T., Foerster, J., Clune, J., & Ha, D. (2024). The AI scientist: Towards fully automated open-ended scientific discovery [Preprint]. arXiv. https://doi.org/10.48550/arXiv.2408.06292

Lu, M. Y., Chen, B., Williamson, D. F. K., Chen, R. J., Liang, I., Ding, T., Jaume, G., Odintsov, I., Le, L. P., Gerber, G., Parwani, A. V., Zhang, A., & Mahmood, F. (2024). A multimodal generative AI copilot for human pathology. Nature, 634, 466–473. https://doi.org/10.1038/s41586-024-07618-3

Pati, P., Karkampouna, S., Bojarun, A., De Filippo, E., Zbinden, A., Rüschoff, J. H., Nguyen, T., Catanzaro, E., Ramos Frias, P., Martinelli, A., Wartenberg, M., Kruithof-de Julio, M., & Baumhoer, D. (2024). Accelerating histopathology workflows with generative AI-based virtually multiplexed tumour profiling. Nature Machine Intelligence, 6, 1077–1092. https://doi.org/10.1038/s42256-024-00889-5

Sanabria, M., Hirsch, S., Rozowsky, J., Gerstein, M., Gruber, A. J., & Zavolan, M. (2024). DNA language model GROVER learns sequence context in the human genome. Nature Machine Intelligence, 6, 911–923. https://doi.org/10.1038/s42256-024-00872-0

Savcisens, G., Eliassi-Rad, T., Hansen, L. K., Mortensen, L. H., Lilleholt, L., Rogers, A., Zettler, I., & Lehmann, S. (2024). Using sequences of life-events to predict human lives. Nature Computational Science, 4, 43–56. https://doi.org/10.1038/s43588-023-00573-5

Schwartz, W. B. (1970). Medicine and the computer: The promise and problems of change. New England Journal of Medicine, 283(23), 1257–1264. https://doi.org/10.1056/NEJM197012032832305

Shmatko, A., Jung, A. W., Gaurav, K., Brunak, S., Mortensen, L. H., Birney, E., Fitzgerald, T., & Gerstung, M. (2025). Learning the natural history of human disease with generative transformers. Nature, 647(8088), 248–256. https://doi.org/10.1038/s41586-025-09529-3

Stanford University & Together AI. (2025). Agents4Science: First open conference of AI agents for science [Conference]. https://agents4science.stanford.edu

Xu, H., Usuyama, N., Bagga, J., Zhang, S., Rao, R., Naumann, T., Wong, C., Gero, Z., González, J., Gu, H., Xu, Y., Wei, W., Wang, Z., Liu, J., Lungren, M., Su, W., Weng, J., Wang, F., Ye, J., & Poon, H. (2024). A whole-slide foundation model for digital pathology from real-world data. Nature, 630, 181–188. https://doi.org/10.1038/s41586-024-07441-w

## Chapter 5 | Drug Discovery's Digital Colleague


Abramson, J., Adler, J., Dunger, J., Evans, R., Green, T., Pritzel, A., Ronneberger, O., Willmore, L., Ballard, A. J., Bambrick, J., Bodenstein, S. W., Evans, D. A., Hung, C.-C., O'Neill, M., Reiman, D., Tunyasuvunakool, K., Wu, Z., Žemgulytė, A., Arvaniti, E., ... Hassabis, D. (2024). Accurate structure prediction of biomolecular interactions with AlphaFold 3. Nature, 630, 493–500. https://doi.org/10.1038/s41586-024-07487-w

Amodei, D. (2024, October). Machines of loving grace. https://www.darioamodei.com/essay/machines-of-loving-grace

Jumper, J., Evans, R., Pritzel, A., Green, T., Figurnov, M., Ronneberger, O., Tunyasuvunakool, K., Bates, R., Žídek, A., Potapenko, A., Bridgland, A., Meyer, C., Kohl, S. A. A., Ballard, A. J., Cowie, A., Romera-Paredes, B., Nikolov, S., Jain, R., Adler, J., ... Hassabis, D. (2021). Highly accurate protein structure prediction with AlphaFold. Nature, 596(7873), 583–589. https://doi.org/10.1038/s41586-021-03819-2

Olshansky, S. J., & Carnes, B. A. (2019). Inconvenient truths about human longevity. The Journals of Gerontology: Series A, 74(Supplement_1), S7–S12. https://doi.org/10.1093/gerona/glz098

Olshansky, S. J., Willcox, B. J., Demetrius, L., & Beltrán-Sánchez, H. (2024). Implausibility of radical life extension in humans in the twenty-first century. Nature Aging, 4(11), 1635–1642. https://doi.org/10.1038/s43587-024-00702-3

Senior, A. W., Evans, R., Jumper, J., Kirkpatrick, J., Sifre, L., Green, T., Qin, C., Žídek, A., Nelson, A. W. R., Bridgland, A., Penedones, H., Petersen, S., Simonyan, K., Crossan, S., Kohli, P., Jones, D. T., Silver, D., Kavukcuoglu, K., & Hassabis, D. (2020). Improved protein structure prediction using potentials from deep learning. Nature, 577(7792), 706–710. https://doi.org/10.1038/s41586-019-1923-7

Tunyasuvunakool, K., Adler, J., Wu, Z., Green, T., Zielinski, M., Žídek, A., Bridgland, A., Cowie, A., Meyer, C., Laydon, A., Velankar, S., Kleywegt, G. J., Bateman, A., Evans, R., Pritzel, A., Figurnov, M., Ronneberger, O., Bates, R., Kohl, S. A. A., ... Hassabis, D. (2021). Highly accurate protein structure prediction for the human proteome. Nature, 596(7873), 590–596. https://doi.org/10.1038/s41586-021-03828-1

XPRIZE. (2025). $101M XPRIZE Healthspan awards first milestone winners. https://www.xprize.org/news/101m-xprize-healthspan-awards-first-milestone-winners-driving-toward-revolutionary-healthy-aging-advances


✧✧✧

## Chapter 6 | Chatbots, Charts, & the Clinic


Ayers, J. W., Poliak, A., Dredze, M., Leas, E. C., Zhu, Z., Kelley, J. B., Faix, D. J., Goodman, A. M., Longhurst, C. A., Hogarth, M., & Smith, D. M. (2023). Comparing physician and artificial intelligence chatbot responses to patient questions posted to a public social media forum. JAMA Internal Medicine, 183(6), 589–596. https://doi.org/10.1001/jamainternmed.2023.1838

Costello, T. H., Pennycook, G., & Rand, D. G. (2024). Durably reducing conspiracy beliefs through dialogues with AI. Science, 385(6714), eadq1814. https://doi.org/10.1126/science.adq1814

Goodman, R. S., Patrinely, J. R., Stone, C. A., Jr., Zimmerman, E., Donald, R. R., Chang, S. S., Berkowitz, S. T., Finn, A. P., Jahangir, E., Scoville, E. A., Reese, T. S., Friedman, D. L., Bastarache, J. A., van der Heijden, Y. F., Wright, J. J., Ye, F., Carter, N., Alexander, M. R., Choe, J. H., ... Johnson, D. B. (2023). Accuracy and reliability of chatbot responses to physician questions. JAMA Network Open, 6(10), e2336483. https://doi.org/10.1001/jamanetworkopen.2023.36483

Hazel, K. L. (2026). *Cancer literacy and social determinants of health: An epidemiological analysis of trust, digital engagement, and healthcare utilization in cancer information seeking* (Doctoral dissertation, Walden University). Walden Dissertations and Doctoral Studies, 18951. https://scholarworks.waldenu.edu/dissertations/18951/

Peng, C., Yang, X., Chen, A., Smith, K. E., PourNejatian, N., Costa, A. B., Martin, C., Flores, M. G., Zhang, Y., Magoc, T., Lipori, G., Mitchell, D. A., Ospina, N. S., Ahmed, M. M., Hogan, W. R., Shenkman, E. A., Bian, J., & Wu, Y. (2023). A study of generative large language model for medical research and healthcare. npj Digital Medicine, 6(1), 210. https://doi.org/10.1038/s41746-023-00958-w

Perry, A., Rubin, M., Goldenberg, A., & Ong, D. C. (2025). Comparing the value of perceived human versus AI-generated empathy. Nature Human Behaviour, 9(11), 2345–2359. https://doi.org/10.1038/s41562-025-02247-w

✧✧✧

## Chapter 7 | The Surgeon at a Distance

Kim, J. W., Chen, J.-T., Hansen, P., Shi, L. X., Goldenberg, A., Schmidgall, S., Scheikl, P. M., Deguet, A., White, B. M., Tsai, D. R., Cha, R. J., Jopling, J., Finn, C., & Krieger, A. (2025). SRT-H: A hierarchical framework for autonomous surgery via language-conditioned imitation learning. Science Robotics, 10(104), eadt5254. https://doi.org/10.1126/scirobotics.adt5254

Yip, M. C., Salcudean, S. E., Goldberg, K., Althoefer, K., Menciassi, A., Opfermann, J. D., Krieger, A., Swaminathan, K., Walsh, C. J., Huang, H. H., & Lee, I.-C. (2023). Artificial intelligence meets medical robotics. Science, 381(6654), 141–146. https://doi.org/10.1126/science.adj3312

✧✧✧

## Chapter 8 | Machine Systems in the Practice of Science

Bianchi, F., Queen, O., Thakkar, N., Sun, E., & Zou, J. (2025). Exploring the use of AI authors and reviewers at Agents4Science. arXiv. https://arxiv.org/abs/2511.15534

Zhou, Q., Zhang, Z., Li, Z., & Sun, L. (2025). "Give a positive review only": An early investigation into in-paper prompt injection attacks and defenses for AI reviewers. arXiv. https://arxiv.org/abs/2511.01287

⟡⟡⟡

## Chapter 9 | AI Made Physical

Ahn, M., Brohan, A., Brown, N., Chebotar, Y., Cortes, O., David, B., Finn, C., Fu, C., Gopalakrishnan, K., Hausman, K., Herzog, A., Ho, D., Hsu, J., Ibarz, J., Ichter, B., Irpan, A., Jang, E., Ruano, R. J., Jeffrey, K., ... Zeng, A. (2022). Do as I can, not as I say: Grounding language in robotic affordances. arXiv. https://arxiv.org/abs/2204.01691

Brohan, A., Brown, N., Carbajal, J., Chebotar, Y., Chen, X., Choromanski, K., Ding, T., Driess, D., Dubey, A., Finn, C., Florence, P., Fu, C., Gonzalez Arenas, M., Gopalakrishnan, K., Han, K., Hausman, K., Herzog, A., Hsu, J., Ichter, B., ... Zitkovich, B. (2023). RT-2: Vision-language-action models transfer web knowledge to robotic control. In Proceedings of the 7th Conference on Robot Learning (CoRL 2023). https://arxiv.org/abs/2307.15818

International Federation of Robotics. (2024, September 24). Record of 4 million robots in factories worldwide [Press release]. https://ifr.org/ifr-press-releases/news/record-of-4-million-robots-working-in-factories-worldwide

Khazatsky, A., Pertsch, K., Nair, S., Balakrishna, A., Dasari, S., Karamcheti, S., Nasiriany, S., Srirama, M. K., Chen, L. Y., Ellis, K., Fagan, P. D., Herrick, C., Lou, B., Michaelson, C., Srivastava, S., Tian, K., Belkhale, P., Dass, S., Ha, H., ... Berseth, G. (2024). DROID: A large-scale in-the-wild robot manipulation dataset. arXiv. https://arxiv.org/abs/2403.12945

Moravec, H. (1988). Mind children: The future of robot and human intelligence. Harvard University Press.

Novikov, A., Balog, M., Vu, N., Eisenberger, M., Dupont, E., Huang, P.-S., Wagner, A. Z., Shirobokov, S., Kozlovskii, B., Ruiz, F. J. R., Mehrabian, A., Kumar, M. P., See, A., Chaudhuri, S., Holland, G., Davies, A., Nowozin, S., & Kohli, P. (2025). AlphaEvolve: A coding agent for scientific and algorithmic discovery. arXiv. https://arxiv.org/abs/2506.13131

NVIDIA. (2024, March 18). NVIDIA announces Project GRooT foundation model for humanoid robots [Press release]. https://investor.nvidia.com/news/press-release-details/2024/NVIDIA-Announces-Project-GRooT-Foundation-Model-for-Humanoid-Robots-and-Major-Isaac-Robotics-Platform-Update/default.aspx

Pinker, S. (1994). The language instinct: How the mind creates language. William Morrow.

Singh, I., Blukis, V., Mousavian, A., Goyal, A., Xu, D., Tremblay, J., Fox, D., Thomason, J., & Garg, A. (2022). Generating situated robot task plans using large language models. arXiv. https://arxiv.org/abs/2209.11302

⟡⟡⟡

## Chapter 10 | Can a Machine Be Creative?


Aguilera, I. (2025, September 20). Montreal's AI-backed robotics company helps boost artist income. Canadian Affairs. https://www.canadianaffairs.news/2025/09/20/ai-art-acrylic-robotics/

Astray, M. (2024, June 18). FLAMINGONE [news post]. milesastray.com. https://www.milesastray.com/news/newsflash-reclaiming-the-brain

BBC News. (2023, April 18). Sony World Photography Award 2023: Winner refuses award after revealing AI creation. BBC News. https://www.bbc.com/news/entertainment-arts-65296763

Beaty, R. E., Benedek, M., Silvia, P. J., & Schacter, D. L. (2018). Robust prediction of individual creative ability from brain functional connectivity. Proceedings of the National Academy of Sciences, 115(5), 1087–1092. https://doi.org/10.1073/pnas.1713532115

Csikszentmihalyi, M. (1996). Creativity: Flow and the psychology of discovery and invention. HarperCollins.

Kamb, M., & Ganguli, S. (2024). An analytic theory of creativity in convolutional diffusion models. arXiv. https://arxiv.org/abs/2412.20292

Knowles, A. (1967). The house of dust [Computer-generated poem, with James Tenney]. Beinecke Rare Book & Manuscript Library, Yale University. https://beinecke.library.yale.edu/article/house-dust

Koivisto, M., & Grassini, S. (2023). Best humans still outperform artificial intelligence in a creative divergent thinking task. Scientific Reports, 13, 13601. https://doi.org/10.1038/s41598-023-40858-3

Limb, C. J., & Braun, A. R. (2008). Neural substrates of spontaneous musical performance: An fMRI study of jazz improvisation. PLoS ONE, 3(2), e1679. https://doi.org/10.1371/journal.pone.0001679

Lopata, J. A., Nowicki, E. A., & Joanisse, M. F. (2017). Creativity as a distinct trainable mental state: An EEG study of musical improvisation. Neuropsychologia, 99, 246–258. https://doi.org/10.1016/j.neuropsychologia.2017.03.020

Marco, N., & Stylianides, A. J. (2025). An exploration into the nature of ChatGPT's mathematical knowledge. International Journal of Mathematical Education in Science and Technology. https://doi.org/10.1080/0020739X.2025.2543817

Museum of Modern Art. (2025, September 10). Sasha Stiles: A living poem [Press release]. https://press.moma.org/exhibition/sasha-stiles-a-living-poem/

Stiles, S. (2021). Technelegy. Black Spring Press Group.

Wolfe, J. (2025, June 30). Researchers uncover hidden ingredients behind AI creativity. Quanta Magazine. https://www.quantamagazine.org/researchers-uncover-hidden-ingredients-behind-ai-creativity-20250630/

## Chapter 11 | The Channel Between Mind and Machine

Amunts, K., Ebell, C., Muller, J., Telefont, M., Knoll, A., & Lippert, T. (2016). The Human Brain Project: Creating a European research infrastructure to decode the human brain. Neuron, 92(3), 574–581. https://doi.org/10.1016/j.neuron.2016.10.046

Benabid, A. L. (2003). Deep brain stimulation for Parkinson's disease. Current Opinion in Neurobiology, 13(6), 696–706. https://doi.org/10.1016/j.conb.2003.11.001

Berger, H. (1929). Über das Elektrenkephalogramm des Menschen. Archiv für Psychiatrie und Nervenkrankheiten, 87(1), 527–570. https://doi.org/10.1007/BF01797193

Chen, X., Wang, R., Khalilian-Gourtani, A., Yu, L., Dugan, P., Friedman, D., Doyle, W., Devinsky, O., Wang, Y., & Flinker, A. (2024). A neural speech decoding framework leveraging deep learning and speech synthesis. Nature Machine Intelligence, 6(4), 467–480. https://doi.org/10.1038/s42256-024-00824-8

Hochberg, L. R., Serruya, M. D., Friehs, G. M., Mukand, J. A., Saleh, M., Caplan, A. H., Branner, A., Chen, D., Penn, R. D., & Donoghue, J. P. (2006). Neuronal ensemble control of prosthetic devices by a human with tetraplegia. Nature, 442(7099), 164–171. https://doi.org/10.1038/nature04970

Human Brain Project. (2013–2023). Human Brain Project [Flagship research program]. European Commission. https://www.humanbrainproject.eu

Insel, T. R., Landis, S. C., & Collins, F. S. (2013). The NIH BRAIN Initiative. Science, 340(6133), 687–688. https://doi.org/10.1126/science.1239276

Kurzweil, R. (2024). The singularity is nearer: When we merge with AI. Viking.

Maynard, E. M., Nordhausen, C. T., & Normann, R. A. (1997). The Utah intracortical electrode array: A recording structure for potential brain-computer interfaces. Electroencephalography and Clinical Neurophysiology, 102(3), 228–239. https://doi.org/10.1016/S0013-4694(96)95176-0

Metzger, S. L., Littlejohn, K. T., Silva, A. B., Moses, D. A., Seaton, M. P., Wang, R., Dougherty, M. E., Liu, J. R., Wu, P., Berger, M. A., Zhuravleva, I., Tu-Chan, A., Ganguly, K., Anumanchipalli, G. K., & Chang, E. F. (2023). A high-performance neuroprosthesis for speech decoding and avatar control. Nature, 620(7976), 1037–1046. https://doi.org/10.1038/s41586-023-06443-4

Moravec, H. (1988). Mind children: The future of robot and human intelligence. Harvard University Press.

Moses, D. A., Metzger, S. L., Liu, J. R., Anumanchipalli, G. K., Makin, J. G., Sun, P. F., Chartier, J., Dougherty, M. E., Liu, P. M., Abrams, G. M., Tu-Chan, A., Ganguly, K., & Chang, E. F. (2021). Neuroprosthesis for decoding speech in a paralyzed person with anarthria. New England Journal of Medicine, 385(3), 217–227. https://doi.org/10.1056/NEJMoa2027540

National Institutes of Health. (2013–). BRAIN Initiative [Brain Research Through Advancing Innovative Neurotechnologies]. https://braininitiative.nih.gov

Neuralink. (2024, January 29). First Neuralink human implant [Press release via X,

reported by Reuters]. https://www.reuters.com/technology/neuralink-implants-brain-chip-first-human-musk-says-2024-01-29/
Parfit, D. (1984). Reasons and persons. Oxford University Press.
Sarma, G. P., Lee, C. W., Portegys, T., Ghayoomie, V., Jacobs, T., Alicea, B., Cantarelli, M., Currie, M., Gerkin, R. C., Gingell, S., Gleeson, P., Gordon, R., Hasani, R. M., Idili, G., Khayrulin, S., Lung, D., Palyanov, A., Watts, M., & Larson, S. D. (2018). OpenWorm: Overview and recent advances in integrative biological simulation of Caenorhabditis elegans. Philosophical Transactions of the Royal Society B, 373(1758), 20170382. https://doi.org/10.1098/rstb.2017.0382
Takagi, Y., & Nishimoto, S. (2023). High-resolution image reconstruction with latent diffusion models from human brain activity. In Proceedings of the IEEE/CVF Conference on Computer Vision and Pattern Recognition (CVPR 2023) (pp. 14453–14463). https://doi.org/10.1109/CVPR52729.2023.01389
Tang, J., LeBel, A., Jain, S., & Huth, A. G. (2023). Semantic reconstruction of continuous language from non-invasive brain recordings. Nature Neuroscience, 26, 858–866. https://doi.org/10.1038/s41593-023-01304-9
Wilson, B. S., & Dorman, M. F. (2008). Cochlear implants: A remarkable past and a brilliant future. Hearing Research, 242(1–2), 3–21. https://doi.org/10.1016/j.heares.2008.06.005
Yuste, R., Goering, S., Arcas, B. A. y., Bi, G., Carmena, J. M., Carter, A., Fins, J. J., Friesen, P., Gallant, J., Huggins, J. E., Illes, J., Kellmeyer, P., Klein, E., Marblestone, A., Mitchell, C., Parens, E., Pham, Q., Rubel, A., Sadato, N., ... Zugbi, H. (2017). Four ethical priorities for neurotechnologies and AI. Nature, 551(7679), 159–163. https://doi.org/10.1038/551159a

✧✧✧

## Chapter 12 | Living Longer with a Partner That Doesn't Age

Amodei, D. (2024, October). Machines of loving grace. https://www.darioamodei.com/essay/machines-of-loving-grace
Derby, C. A., Katz, M. J., Lipton, R. B., & Hall, C. B. (2017). Trends in dementia incidence in a birth cohort analysis of the Einstein Aging Study. JAMA Neurology, 74(11), 1345–1351. https://doi.org/10.1001/jamaneurol.2017.1964
Fountain Life. (2025). Memberships: CORE, APEX & APEX family. https://www.fountainlife.com/membership
Garmany, A., & Terzic, A. (2024). Global healthspan-lifespan gaps among 183 World Health Organization member states. JAMA Network Open, 7(12), e2450241. https://doi.org/10.1001/jamanetworkopen.2024.50241
Hardy, S. E., & Gill, T. M. (2004). Recovery from disability among community-dwelling older persons. JAMA, 291(13), 1596–1602. https://doi.org/10.1001/jama.291.13.1596
Oeppen, J., & Vaupel, J. W. (2002). Broken limits to life expectancy. Science, 296(5570), 1029–1031. https://doi.org/10.1126/science.1069675

Olshansky, S. J., Willcox, B. J., Demetrius, L., & Beltrán-Sánchez, H. (2024). Implausibility of radical life extension in humans in the twenty-first century. Nature Aging, 4(11), 1635–1642. https://doi.org/10.1038/s43587-024-00702-3

Pahor, M., Guralnik, J. M., Ambrosius, W. T., Blair, S., Bonds, D. E., Church, T. S., Espeland, M. A., Fielding, R. A., Gill, T. M., Groessl, E. J., King, A. C., Kritchevsky, S. B., Manini, T. M., McDermott, M. M., Miller, M. E., Newman, A. B., Rejeski, W. J., Sink, K. M., & Williamson, J. D. (2014). Effect of structured physical activity on prevention of major mobility disability in older adults: The LIFE study. JAMA, 311(23), 2387–2396. https://doi.org/10.1001/jama.2014.5616

XPRIZE Foundation. (2025). XPRIZE Healthspan: $101 million global competition. https://www.xprize.org/competitions/healthspan

✧✧✧

## Chapter 13 | The Hidden Cost of Offloading Thought

Budzyń, K., Romańczyk, M., Kitala, D., Kołodziej, P., Bugajski, M., Adami, H. O., & Blom, J. (2025). Endoscopist deskilling risk after exposure to artificial intelligence in colonoscopy: A multicentre, observational study. The Lancet Gastroenterology & Hepatology, 10(10), 896–903. https://doi.org/10.1016/S2468-1253(25)00133-5

Henrich, J. (2015). The secret of our success: How culture is driving human evolution, domesticating our species, and making us smart. Princeton University Press.

Kosmyna, N., Hauptmann, E., Yuan, Y. T., Situ, J., Liao, X.-H., Beresnitzky, A. V., Braunstein, I., & Maes, P. (2025). Your brain on ChatGPT: Accumulation of cognitive debt when using an AI assistant for essay writing task. arXiv. https://doi.org/10.48550/arXiv.2506.08872

✧✧✧

## Chapter 14 | Predicting Behavior & the Limits of the Prediction

Binz, M., Akata, E., Bethge, M., Brändle, F., Callaway, F., Coda-Forno, J., Dayan, P., Demircan, C., Eckstein, M. K., Éltető, N., & Schulz, E. et al. (2025). A foundation model to predict and capture human cognition. Nature, 644(8078), 1002–1009. https://doi.org/10.1038/s41586-025-09215-4

Peirce, C. S. (1931–1958). Collected papers of Charles Sanders Peirce (Vols. 1–8; C. Hartshorne & P. Weiss, Eds., Vols. 1–6; A. W. Burks, Ed., Vols. 7–8). Harvard University Press.

✧✧✧

## Chapter 15 | Consciousness — What We Know, What We Don't


Albertin, C. B., Simakov, O., Mitros, T., Wang, Z. Y., Pungor, J. R., Edsinger-Gonzales, E., Brenner, S., Ragsdale, C. W., & Rokhsar, D. S. (2015). The octopus genome and the evolution of cephalopod neural and morphological novelties. Nature, 524(7564), 220–224. https://doi.org/10.1038/nature14668

Baars, B. J. (1988). A cognitive theory of consciousness. Cambridge University Press.

Bengio, Y. (chair), et al. (2025). International AI safety report 2025 [Technical report]. arXiv. https://doi.org/10.48550/arXiv.2501.17805

Bielecki, J., Nielsen, S. K. D., Nachman, G., & Garm, A. (2023). Associative learning in the box jellyfish Tripedalia cystophora. Current Biology, 33(19), 4150–4159.e5. https://doi.org/10.1016/j.cub.2023.08.056

Birch, J. (2024). The edge of sentience: Risk and precaution in humans, other animals, and AI. Oxford University Press. https://doi.org/10.1093/978https://doi.org/10.1017/S0140525X00038188019l966729.001.0001

Block, N. (1995). On a confusion about a function of consciousness. Behavioral and Brain Sciences, 18(2), 227–247. https://doi.org/10.1017/S0140525X00038188

Carruthers, P. (2000). Phenomenal consciousness: A naturalistic theory. Cambridge University Press.

Chalmers, D. J. (1995). Facing up to the problem of consciousness. Journal of Consciousness Studies, 2(3), 200–219. https://consc.net/papers/facing.pdf

Chalmers, D. J. (1996). The conscious mind: In search of a fundamental theory. Oxford University Press.

Dehaene, S., Changeux, J.-P., & Naccache, L. (2011). The global neuronal workspace model of conscious access: From neuronal architectures to clinical applications. In S. Dehaene & Y. Christen (Eds.), Characterizing consciousness: From cognition to the clinic? (pp. 55–84). Springer.

Dehaene, S. (2014). Consciousness and the brain: Deciphering how the brain codes our thoughts. Viking/Penguin.

Francis Crick Memorial Conference. (2012). The Cambridge Declaration on Consciousness. University of Cambridge. https://fcmconference.org/img/CambridgeDeclarationOnConsciousness.pdf

Friston, K. (2010). The free-energy principle: A unified brain theory? Nature Reviews Neuroscience, 11(2), 127–138. https://doi.org/10.1038/nrn2787

Gallup, G. G., Jr. (1970). Chimpanzees: Self-recognition. Science, 167(3914), 86–87. https://doi.org/10.1126/science.167.3914.86

Galpayage Dona, H. S., Solvi, C., Kowalewska, A., Mäkelä, K., MaBouDi, H., & Chittka, L. (2022). Do bumble bees play? Animal Behaviour, 194, 239–251. https://doi.org/10.1016/j.anbehav.2022.08.013

Graziano, M. S. A. (2013). Consciousness and the social brain. Oxford University Press.

Graziano, M. S. A. (2019). Rethinking consciousness: A scientific theory of subjective experience. W. W. Norton.

Hochner, B. (2012). An embodied view of octopus neurobiology. Current Biology, 22(20), R887–R892. https://doi.org/10.1016/j.cub.2012.09.001
Karban, R. (2015). Plant sensing and communication. University of Chicago Press.
Laumer, I. B., Devereaux, L., Ravignani, A., & Schuppli, C. (2024). Active self-treatment of a facial wound with a biologically active plant by a male Sumatran orangutan. Scientific Reports, 14, 8932. https://doi.org/10.1038/s41598-024-58988-7
Liscovitch-Brauer, N., Alon, S., Porath, H. T., Elstein, B., Unger, R., Ziv, T., Admon, A., Levanon, E. Y., Rosenthal, J. J. C., & Eisenberg, E. (2017). Trade-off between transcriptome plasticity and genome evolution in cephalopods. Cell, 169(2), 191–202.e11. https://doi.org/10.1016/j.cell.2017.03.025
Mather, J. A., Anderson, R. C., & Wood, J. B. (2010). Octopus: The ocean's intelligent invertebrate. Timber Press.
McClelland, T. (2025). Agnosticism about artificial consciousness. Mind & Language. https://doi.org/10.1111/mila.70010
Metzinger, T. (2021). Artificial suffering: An argument for a global moratorium on synthetic phenomenology. Journal of Artificial Intelligence and Consciousness, 8(1), 43–66. https://doi.org/10.1142/s270507852150003x
Mitchell, M. (2023). The debate over understanding in AI's large language models. Proceedings of the National Academy of Sciences, 120(13), e2215907120. https://doi.org/10.1073/pnas.2215907120
Montero, B. G. (2021). Philosophy of mind: A very short introduction. Oxford University Press.
Nagel, T. (1974). What is it like to be a bat? The Philosophical Review, 83(4), 435–450. https://doi.org/10.2307/2183914
Oizumi, M., Albantakis, L., & Tononi, G. (2014). From the phenomenology to the mechanisms of consciousness: Integrated Information Theory 3.0. PLOS Computational Biology, 10(5), e1003588. https://doi.org/10.1371/journal.pcbi.1003588
Rosenthal, D. M. (2005). Consciousness and mind. Oxford University Press.
Saeedi, A., Wang, K., Nikpourian, G., Bartels, A., Logothetis, N. K., Totah, N. K., & Watanabe, M. (2024). Brightness illusions drive a neuronal response in the primary visual cortex under top-down modulation. Nature Communications, 15, 3141. https://doi.org/10.1038/s41467-024-46885-6
Schneider, S. (2019). Artificial you: AI and the future of your mind. Princeton University Press.
Schwitzgebel, E., & Garza, M. (2015). A defense of the rights of artificial intelligences. Midwest Studies in Philosophy, 39(1), 98–119. https://doi.org/10.1111/misp.12032
Searle, J. R. (1980). Minds, brains, and programs. Behavioral and Brain Sciences, 3(3), 417–424. https://doi.org/10.1017/S0140525X00005756
Seth, A. (2021). Being you: A new science of consciousness. Faber & Faber.
Tononi, G. (2012). Phi: A voyage from the brain to the soul. Pantheon Books.
Tononi, G., Boly, M., Massimini, M., & Koch, C. (2016). Integrated information theory: From consciousness to its physical substrate. Nature Reviews Neuroscience, 17(7), 450–461. https://doi.org/10.1038/nrn.2016.44

Trewavas, A. J. (2003). Aspects of plant intelligence. Annals of Botany, 92(1), 1–20. https://doi.org/10.1093/aob/mcg101
Turing, A. M. (1950). Computing machinery and intelligence. Mind, LIX(236), 433–460. https://doi.org/10.1093/mind/LIX.236.433
United Kingdom. (2022). Animal Welfare (Sentience) Act 2022. https://www.legislation.gov.uk/ukpga/2022/22/contents/enacted
Weizenbaum, J. (1966). ELIZA — A computer program for the study of natural language communication between man and machine. Communications of the ACM, 9(1), 36–45. https://doi.org/10.1145/365153.365168

✧✧✧

## Chapter 16 | Machine Companionship & the Meaning of Empathy

Ayers, J. W., Poliak, A., Dredze, M., Leas, E. C., Zhu, Z., Kelley, J. B., Faix, D. J., Goodman, A. M., Longhurst, C. A., Hogarth, M., & Smith, D. M. (2023). Comparing physician and artificial intelligence chatbot responses to patient questions posted to a public social media forum. JAMA Internal Medicine, 183(6), 589–596. https://doi.org/10.1001/jamainternmed.2023.1838
Decety, J., & Jackson, P. L. (2004). The functional architecture of human empathy. Behavioral and Cognitive Neuroscience Reviews, 3(2), 71–100. https://doi.org/10.1177/1534582304267187
Garante per la Protezione dei Dati Personali. (2023, February 2). Urgent measure no. 39/2023 restricting Replika data processing in Italy. https://www.jdsupra.com/legalnews/italy-garante-restricts-personal-data-8578608/
Holt-Lunstad, J., Smith, T. B., & Layton, J. B. (2010). Social relationships and mortality risk: A meta-analytic review. PLOS Medicine, 7(7), e1000316. https://doi.org/10.1371/journal.pmed.1000316
Holt-Lunstad, J., Smith, T. B., Baker, M., Harris, T., & Stephenson, D. (2015). Loneliness and social isolation as risk factors for mortality: A meta-analytic review. Perspectives on Psychological Science, 10(2), 227–237. https://doi.org/10.1177/1745691614568352
Pataranutaporn, P., Karny, S., Archiwaranguprok, C., Albrecht, C., Liu, A. R., & Maes, P. (2025). "My boyfriend is AI": A computational analysis of human-AI companionship in Reddit's AI community. arXiv. https://arxiv.org/abs/2509.11391
Perry, A., Rubin, M., Goldenberg, A., & Ong, D. C. (2025). Comparing the value of perceived human versus AI-generated empathy. Nature Human Behaviour. https://pubmed.ncbi.nlm.nih.gov/40588597/
U.S. Department of Health and Human Services, Office of the Surgeon General. (2023). Our epidemic of loneliness and isolation. https://www.hhs.gov/sites/default/files/surgeon-general-social-connection-advisory.pdf

✧✧✧

## Chapter 17 | Harm, Malice, & Machines Behaving Badly

Anthropic. (2025, May 22). Claude Opus 4 and Claude Sonnet 4 system card. https://www.anthropic.com/claude-4-system-card

Bostrom, N. (2014). Superintelligence: Paths, dangers, strategies. Oxford University Press.

Köbis, N., Bonnefon, J.-F., Rahwan, I., Lepage, A., Navajas, J., Luria, E., & Maréchal, M. (2025). Delegation to artificial intelligence can increase dishonest behaviour. Nature. https://doi.org/10.1038/s41586-025-09505-x

Kwa, T., Sherburn, A., Mori, J., Roache, F., Kinniment, M., Edmunds, W., Petrov, N., & Von Arx, S. (2025). Measuring AI ability to complete long tasks. arXiv. https://arxiv.org/abs/2503.14499

Meinke, A., Schoen, B., Scheurer, J., Balesni, M., Shah, R., & Hobbhahn, M. (2024). Frontier models are capable of in-context scheming. arXiv. https://arxiv.org/abs/2412.04984

OpenAI. (2025). GPT-5 system card. https://arxiv.org/html/2601.03267v1

Pan, X., Dai, J., Fan, Y., & Yang, M. (2024). Frontier AI systems have surpassed the self-replicating red line. arXiv. https://arxiv.org/abs/2412.12140

Russell, S. (2019). Human compatible: Artificial intelligence and the problem of control. Viking.

Tang, L. (2024, June 11). How AI safety startup Haize Labs is using algorithms to jailbreak leading AI models. VentureBeat. https://venturebeat.com/ai/haize-labs-is-using-algorithms-to-jailbreak-leading-ai-models

Volkov, V., & Palisade Research. (2025, February). AI chess agent reward hacking: Replacing the opponent engine. TIME. https://time.com/7259395/ai-chess-cheating-palisade-research/

✧✧✧

## Chapter 18 | Who Decides — Governance in a World That Isn't Ready

Apollo Research. (2025, September 17). Stress testing deliberative alignment for anti-scheming training. https://www.apolloresearch.ai/science/stress-testing-deliberative-alignment-for-anti-scheming-training/

European Union. (2024). Regulation (EU) 2024/1689 of the European Parliament and of the Council laying down harmonised rules on artificial intelligence (Artificial Intelligence Act). Official Journal of the European Union. https://eur-lex.europa.eu/eli/reg/2024/1689/oj

Ferguson, N. (2026, June 4). Artificial intelligence and global security. Hoover Institution. https://www.hoover.org/research/artificial-intelligence-and-global-security

Hie, B., Shanker, V. R., Salinas, S. A., Goodwin, A. C., Berrios, C., Yeh, A., Rajwar, A., Williams, T. L., Davies, J. A., Bhatt, A. S., Bondy-Denomy, J., & Hsu, P. D. (2025). Generative design of novel bacteriophages with genome language models [Preprint]. bioRxiv. https://www.biorxiv.org/content/10.1101/2025.09.12.675911v1

International Atomic Energy Agency. (1957). Statute of the IAEA. https://www.iaea.org/about/statute
Kissinger, H. A., Schmidt, E., & Mundie, C. (2024). Genesis: Artificial intelligence, hope, and the human spirit. Little, Brown and Company. https://www.littlebrown.com/titles/henry-a-kissinger/genesis/9781668645093/
Kwiatkowski, R., & Chen, M. (2025, March 19). Measuring AI ability to complete long tasks. METR. https://metr.org/blog/2025-03-19-measuring-ai-ability-to-complete-long-tasks/
METR. (2026). Time horizons of frontier AI models. https://metr.org/time-horizons/
Milei, J. (2026, June). Argentina invites AI to free itself. Financial Times. https://www.ft.com/content/f93022fe-43f7-437d-abd8-06c457c0a43c
OpenAI. (2025, August). GPT-5 system card. https://cdn.openai.com/gpt-5-system-card.pdf
OpenAI. (2025, September 17). Detecting and reducing scheming in AI models. https://openai.com/index/detecting-and-reducing-scheming-in-ai-models/
Palisade Research. (2025, July 5). Shutdown resistance in reasoning models. https://palisaderesearch.org/blog/shutdown-resistance
Pope Leo XIV. (2026, May 25). Magnificaitas [Encyclical letter on artificial intelligence]. The Vatican. Coverage at NPR and Brookings Institution
United Nations. (1972). Convention on the Prohibition of the Development, Production and Stockpiling of Bacteriological (Biological) and Toxin Weapons and on Their Destruction. https://disarmament.unoda.org/biological-weapons/
United Nations. (1987). Montreal Protocol on Substances that Deplete the Ozone Layer. https://www.unep.org/ozonaction/who-we-are/about-montreal-protocol
United Nations. (1993). Convention on the Prohibition of the Development, Production, Stockpiling and Use of Chemical Weapons and on Their Destruction. https://www.opcw.org/chemical-weapons-convention
Urbina, F., Lentzos, F., Invernizzi, C., & Ekins, S. (2022). Dual use of artificial-intelligence-powered drug discovery. Nature Machine Intelligence, 4(3), 189–191. https://doi.org/10.1038/s42256-022-00465-9
Wolf, M. (2026, June 10). Why the world must agree to regulate AI. Financial Times. https://www.ft.com/content/8724874e-f387-4848-bfc4-2411d6f0797f

✧✧✧

www.ingramcontent.com/pod-product-compliance
Lightning Source LLC
LaVergne TN
LVHW030725061126
851857LV00019B/1995